I0467535

# When Disaster Strikes

## How Healthcare Facilities respond

**Copyright © 2016 by Lee G. Shanley**

**All rights reserved**

**Cover Art by Jessica H. Granger**

**All rights reserved**

# Foreword

Much has been written over the years regarding all aspects of Emergency Management. But specific information regarding hospitals and healthcare facilities is much less prolific. To that end, it is my intent to present in this book, two different aspects of hospital and healthcare Emergency Management. The first part of this book will deal specifically with preparation, prevention, mitigation and response to community based disasters. How and why facilities need to build a network of resources, equipment and mutual aid will be illustrated and discussed. This became all too clear, especially in the Northeast, after super storm Sandy devastated the eastern seaboard in 2012.

Part two of this book however, will deal with something that has had very little written about it, but is something that keeps all healthcare Emergency Managers up at night. What happens when the healthcare facility is the site of the disaster? Who helps them? How do they treat victims? How do they protect staff and patients? How do they maintain basic functions? And, maybe most important of all, how do they prevent the first responders from becoming potential secondary victims themselves? These are just some of the questions that will be addressed within these pages.

Hopefully, this book will provide the reader with some of the knowledge that will be needed to successfully prepare for, respond to and recover from, a disaster in their community or at their own healthcare facility. If nothing else, for the healthcare Emergency Manager, it should provide enough food for thought to start asking the most important question of all. ***Are we prepared?***

# Dedication

This book is dedicated to *Mary Theresa*, my wife, without whose inspiration, love and support, this book would never have been started, let alone brought to fruition.

**"We are not put on this earth for ourselves, but are placed here for each other. If you are there always for others, then in time of need, someone will be there for you."**

**Jeff Warner**

# TABLE OF CONTENTS

# PART I

# When the event strikes the community

# CHAPTER 1

## What are we preparing for?

Much of what is presented in chapter one may seem very basic, Emergency Management 101, if you will. Many of these things you may have already viewed in other books or courses. But for the new practitioner, it's a good foundation and for the seasoned veteran, it's a good review. One of the first questions we need to ask ourselves as Emergency Managers, is what exactly are we preparing our Healthcare facilities for? And conversely, what are we protecting them from? Once we've determined the answer to those questions, everything else follows. The answers will be different for all facilities as well as all Emergency Managers.

For instance, when we look at weather related disasters, we can generalize and say that the Northeast prepares for hurricanes and winter storms, while the Midwest prepare for tornadoes and drought and the West prepares for earthquakes and wildfires. But this is really just the tip of the iceberg. Much of the impact from natural weather related disasters is directly impacted by topography as well as density of population.

The cost of a disaster can also be tied directly to property and density of population as well. Generally, the more populated an area is, the higher degree of economic damage will occur. Property values also have an impact on the economic cost.

Maybe one of the very first things we need to discuss is, how you define a disaster. What is the difference between an emergency and a disaster? Is every emergency a disaster and vice versa? Let's start with a workable definition of a disaster.

*"A disaster is an emergency or incident of such severity, magnitude and duration, that it completely overwhelms the facility, its staffing and resources, preventing it from operating in a normal or routine mode".*

Therefore, the busted water pipe that's floods a unit, a small trash fire with a lot of smoke, a partial power outage, a water shutdown, having to evacuate a unit for a specific reasons, even a fairly large medical surge are good examples of emergencies but not necessarily disasters. These are the things we deal with in the normal course of a day's work.

*While these types of incidents tax our resources, they don't overwhelm them. And that's the key difference.*

The following is very basic information, but does need to be addressed. Let us then look at how disasters are categorized. Researchers have traditionally defined three different categories of disasters; *Natural, Technological and Human Caused (formerly civil) disasters.*

*Natural disasters* are then broken down into three sub categories;

*Climatological*; events such as tornadoes, floods, hurricanes, wildfires and severe winter storms are good examples of these.

*Geophysical; these are derived from a sudden and violent shift in the earth's surface.*

*Health Hazards & Epidemics; some examples are, severe influenza, viral infections, and food stock contaminations.*

Events such as volcanic eruptions, Tsunami's, large landslides, and of course Earthquakes represent this category of disasters.

Another example that has become much more prevalent around the country in recent years is Sinkholes. We've seen a lot of these types of occurrences in Florida.

The effects of Climatological and Geophysical disasters are well known and have been documented thoroughly. There have even been specific response plans written for healthcare facilities to deal with them. So we're fairly comfortable with our response plans.

That is, unless the healthcare facility is directly impacted by the disaster like Tropical Storm Allison in 2001 Houston, TX, (M. D. Anderson, Baylor Health Science Center).

Or hurricane Katrina in 2005 did to New Orleans (Charity Hospital), the 2011 tornado in Joplin, Missouri (St. John's Mercy Regional Medical Center), hurricane Sandy in 2012 in New York (NYU Bellevue and Long Beach Hospital) and the 2013 tornado did in Moore, Oklahoma (Moore Medical Center)to name but a few examples. Then it becomes a completely different story, with a different focus and response. This is exactly the type of thing we'll discuss in more detail in Part II of this book. What happens when the facility is the disaster focal point?

What we're not as comfortable with surprisingly, since we are healthcare facilities, is our response to the following category of Natural disasters. Specifically, health hazards and pandemics.

*Health Hazards; Pandemics,* Our country because of its very size and diversity is open to potential health disasters. Looking back through the years, the influenza pandemics of 1918, 1957 and 1967 killed millions not only in this country but around the world. On average, the flu kills about 25,000 people a year in the United States.

This is according to the 'Centers for Disease Control" (CDC), though the range varies greatly each year depending upon the strain and virility of the virus. Most deaths are actually caused by complications from the flu. Add to that, the West Nile virus, SARS (Sudden Acute Respiratory Syndrome), H5N1 (Avian Flu) H1N1 (Swine flu), Enterovirus D68 which has caused polio like symptoms in some children. Also, the recent outbreak of plague in Madagascar and the Ebola epidemic in West Africa, the outbreak of Measles in this country, the rise of tuberculosis cases and now Zika virus and you have just some idea of the potential health disasters that can affect you, your community and your facility.

The 1918-1919 Flu Pandemic killed more people worldwide (approximately 30 million) than World War I (approximately 17 million). And this was accomplished in a much shorter timeframe. Two years as opposed to four years. It spread across the United States in about eight weeks. This was before commercial air travel, when most people still travelled by train or horse. The vast majority of the population still did not own cars. Yet it still spread quickly across the country fairly rapidly. People's lives were changed drastically resulting from measures that towns and cities put in place such as quarantines. Can you even imagine how fast a virus of that strength and nature could and would spread in today's day and age?

How would we respond; locally, regionally and nationally. What actions could we take to protect our staff and how would we be able to function in response to the anticipated surge of patients that we would receive.

In 1918 most people did not go to hospitals when they were sick, and yet the entire national healthcare infrastructure was still overwhelmed. Facilities ran out of healthcare providers, as they in turn also became ill.

It was estimated that approximately twenty-five per cent of the healthcare workforce were incapacitated, (ill or dead) from the flu.

As a result, allied medical titles such as, medical students, nursing students even Dentists, in some cases were authorized for medical service. Retired doctors and nurses were also called back to serve. Today, it's almost one hundred years later. What would happen now in a similar situation? Do you think we would fare any better today with all our advances in healthcare? It's true, we have more knowledge, better equipment, better trained staff, better medications and resources. But more people today use hospitals as their primary medical provider. How quickly would your facility become stressed? How comfortable are you with our Local, State, Regional and National Healthcare Infrastructure? As Healthcare facilities, how would you function during a national healthcare crisis? Would you and could you institute *altered standards of care*? What facility right now is willing to treat some victims and let others die? That's battlefield medicine at its essence. Maybe the most important question of all, could you and your facility survive?

Personally, I feel we can and could handle a local, regional or even cross jurisdictional health crisis successfully, pretty much anywhere in the country. However, in my opinion, if it was a nationwide crisis, the response and results would be much different. Even with all of the focus and attention that has been forthcoming from the Federal Government, I still don't believe that the national healthcare infrastructure is set up to handle mass victims. I feel the same about mass casualties. In theory the framework is there and we should be able to handle it.  In reality, I don't believe we have the proper mindset across the country to deal with this type of situation cohesively. What if states that were better prepared, or had more supplies and medications decided to treat their residents only knowing no further supplies would be forthcoming. Would we have "border wars"? I believe in the end, we would address it, but it would be much more disjointed functionally. It might take something like a "Medical Marshal Law" to bring it all together.

An interesting fact about the 1918-1919 Influenza Pandemic is that in certain cities like St. Louis, Missouri, the spread of the disease and subsequent fatality rates were significantly lower than that of many other cities around the country.

This was a direct result of them instituting extreme isolation and quarantine measures. In essence, they declared that "Medical Marshal Law" I just spoke about. It's something that not only worked very well for them but saved countless numbers of lives. Could we and would we institute something like that in today's day and age? Remember, this was 1918, the public at large was still fairly docile and more often than not followed what the government said without question. I'm not sure that people today would be as accepting in having their personal rights and freedoms infringed upon even if it was for their own health and welfare. What would be the alternative; rioting, civil unrest or even a breakdown of society, as fear would overcome common sense.

Let's now look at something that has the potential for being equally frightening. As healthcare providers, we also need to be concerned with and prepared for disasters that can impact our *food chain and agricultural industry*. In 2014 for instance, there was a massive recall of almost two million pounds of beef as a result of *E-coli* contamination. E-coli outbreaks have also been linked and traced to spinach, lettuce as well as raw milk. Products we use every day. We've also experienced outbreaks of *Salmonella* in our food chain in 2010, 2012 and 2013. These have been traced primarily to our egg and poultry industry. There was a *Listeria* outbreak traced to an ice cream manufacturer in 2015. And as a final example, we have *Mad-cow disease*, primarily affecting the beef industry.

A global example of this instance would be the United Kingdom which suffered an outbreak in 1993 that infected some 180,000 cattle with almost 4.4 million being slaughtered. As you can see, there are potential threats everywhere. These things are here to stay so we need to be aware of them, weigh their potential risks to your facility and prepare your response to the best of your ability against them.

In this book, we'll discuss what our response options are regarding some of these disasters and how we can protect and treat not only out patients but maybe more importantly, our staff. Remember, staff are our greatest asset. For without them, we cannot perform our primary function. If that happens, then we cease to be an asset to our community. It is my sincere hope, that this book will be able to provide some direction and insight regarding these issues as we delve deeper into the question of how do we prepare for disasters.

The next category of disasters we'll talk about are *Technological Disasters*. These are events generally caused by actions of human intervention through neglect or omission. They tend not to be intentional but rather accidental or inadvertent in nature. Some classic examples of this category would be the British Petroleum (BP) Deep Horizon explosion, fire and subsequent oil spill of nearly 250 million gallons in the Gulf of Mexico in 2010.

The Exxon Valdez oil tanker in 1989, ran aground spilling over 10 million gallons of crude in Alaska. The Union Carbide disaster of 1984 in Bhopal India in which a valve was inadvertently left open, which subsequently resulted in the release of a toxic cloud of *methyl isocyanate gas* that exposed over 500,000 people and killed almost 4,000 people.

None of these disasters were intentional, but they were the direct result of human intervention or negligence. The most common of all technological disasters are fires and explosions. These are followed closely by transportation incidents.

For example; train derailments have a significant impact on our communities given the hazardous cargo they sometimes carry. The fact that many times these trains go directly through population centers poses a constant hazard concern.

Unlike many major metropolitan centers, many train routes take them directly through Middle America. This alone, puts many thousands of people at risk. One example of this is the 2015 crude oil derailment in Boomer, West Virginia where 14 tank cars of a 100 car train derailed and burst into flames. Two local towns were evacuated and the spill may have greatly impacted the local water supply. The 2014 crude oil train derailment in Lynchburg Virginia and the 2013 Lac Megantic, Quebec derailment which left 47 people dead are just some recent examples. In all three cases there were subsequent explosions and fires. Each one closed much of the local towns down temporarily.

Almost half of Lac Megantic was destroyed. This then brings in the negative economic aspect pertaining to disasters. As I stated earlier, technological disasters are not generally intentional in nature, but that does not take away from the fact that they can have devastating results.

Healthcare facilities need to be aware of all of these potential hazards in their community and plan accordingly for a response, should that be necessary.

There is one disaster event which I like to address briefly here and mention that it actually bridges a gap between Technological disasters and Human Caused Disasters. I'm referring to *Cyber Events/Terrorism*. Cyber events have increased dramatically over the last several years both domestically and internationally. It impacts consumers, businesses (think large retailers) and even governments. It has most certainly impacted healthcare facilities.

In today's day and age almost all businesses are computer run and managed. Nowhere is this truer than for Hospitals and Healthcare facilities. From admission, to physical & history, doctor's orders, treatment plan, medication, insurance and finally discharge, the patient's life is tied directly to some kind of computer program. Should that system be compromised, hacked, shut down or crash the results could be catastrophic. Most facilities today do not have a paper back-up system or staff that know how to implement it. Even if they do, it's usually very limited in nature.

This is an area where the Emergency Manager, Information Technology (I.T.) Director and Administration need to work closely together. This will assure that there is a continuity of operations plan in place, with built in redundancy, should the main system fail.

The United States Department of Homeland Security has a publication titled; *"Blueprint for a Secure Cyber Future"* that is available online to assist you.

Just recently, in an article published and picked up around the country, Admiral Michael Rogers, director of the National Security Agency (NSA) and head of U.S. Cyber Command stated the following. "China and one or two other countries are capable of mounting cyber-attacks that would shut down the electric grid and other critical systems in parts of the United States. There is also a very well written and informative book on this very subject titled "Lights Out" by author Ted Koppel". All of this leads me to believe it is only a matter of when, not if, we are going to see something dramatic". This is indeed a very sobering thought.

The third category of disasters we'll address is *Human Caused Disasters* (formerly referred to as Civil Disasters). The major difference between these and technological disasters is that the human intervention aspect for these *is intentional.*

When we refer to Human Caused Disasters we're generally speaking about large riots, massed protests turning violent, terrorist acts and even war. Human Caused Disasters tend to arrive with little or no warning and can be totally unexpected and unpredictable in nature. A single incident, in some cases, can set off a massive, violent response.

Take for example, the 1992 Los Angeles riots, sometimes referred to as the Rodney King riots. In this case, over a period of six days, from April 29th to May 4th more than 11,000 people were arrested. In addition, over 4,000 injured and 53 people were left dead. Hospital Emergency Departments became battlefield triage centers.

The police who were totally overwhelmed, received military support from the National Guard, U.S. Army and the U.S. Marines who were called in. Estimates were that there was over one billion dollars' worth of property damage just in the Los Angeles area alone. The resulting economic impact on the surrounding areas, state and nation were never fully tallied.

In regard to war, we've been extremely fortunate. Not since the Civil War has there been a war fought on mainland American soil. But in other parts of the world, it seems to be an everyday occurrence. Just look at Egypt, Iraq or Syria. And it's not just the Middle-East. Now in Eastern Europe, there is constant fighting and turmoil. And in Africa, many nations are under siege.

The effects of war are disastrous on a country, completely devastating the infrastructure. One of the first things to be impacted and in some cases targeted are healthcare facilities.

On the other hand, terrorism, unfortunately, is all too familiar to us now. As it is to the rest of the world as well. First and foremost, let us try to define Terrorism. This is not a simple or straight forward task by any means. Currently, there is no legal International consensus regarding the definition of Terrorism. Nor is there an Academic one. One source, *Schmid and Jongman 1988* counted 109 definitions of terrorism.

"Terrorism expert *Walter LaQueur* also has counted over 100 definitions and concludes that the *"only general characteristic generally agreed upon is that terrorism involves violence and the threat of violence."*

In this book, for simplification and some clarity, we'll use the following F.B.I. definition of terrorism;

*"The unlawful use of violence and intimidation against people or property to coerce or intimidate in the pursuit of political, religious or ideological aims."*

Terrorism can be domestic or international in nature. There is also a third term that is sometimes applied to terrorist acts and that is transnational terrorism.

Domestic terrorism takes place against American citizens on U.S. soil. International or State sponsored terrorism is financially and strategically run by foreign governments against American citizens. Transnational terrorism is organized by separate groups or individuals espousing specific religious or ideological goals. These kind of attacks can occur against Americans anywhere in the world.

Unfortunately, terrorism has become a staple of the twentieth and now the twenty-first century. The following are several examples of each terrorism classification.

Beginning with the 1996 bombing of Centennial Park at the Atlanta Olympic Games in which an individual "Eric Rudolf" detonated a bomb that killed one person and injured 111 others. Then we have an individual who committed domestic terrorism across this country for a period of almost twenty years, from the years 1978-1995 "Ted Kaczynski"; also known as the *"Unabomber"*. He was eventually turned into authorities by a family member.

Also included as an example of domestic terrorism is the 1995 Oklahoma City bombing of the Murrah Federal Building. The force of the blast which killed indiscriminately, took the lives of some 168 people and injured almost 700. It damaged or destroyed 324 buildings within a 16 block radius of the blast site. The estimated damage was over 650 million dollars.

In this case all healthcare facilities very quickly became overwhelmed in the initial phases. But by having plans in place and working together with Law Enforcement, EMS and the community they stayed a viable force throughout the ordeal.

As far as *International Terrorism* we have the Korean Air Lines bombing of flight 858 in 1987. One hundred and fifteen people died. The bombing was tied to the 50th anniversary of the armistice between North and South Korea. There was also the Pan Am flight 103 bombing which took place over Lockerbie, Scotland in 1988. That incident killed 270 people and was later traced by the F.B.I. back to Libya and Moamar Khadaffi.

Depending on your beliefs and or political leaning, one could even argue that the 200 years of the "Crusades" were "State sponsored terrorism" on both sides. And what about the almost 400 hundred years of the "Inquisition".

 Some people are comparing some of their methods like "forced conversions" to what the Islamic State of Iraq and Syria (ISIL/ISIS) is now doing in the Middle East.

 And finally, probably the worst ever state sponsored terrorism reign was that of the National Socialist or Nazi Party, where from 1933-1945 it is estimated that over 50 million people died as a direct result of them worldwide. (Military & civilian)

The third piece in this triangle is that of *Transnational Terrorism*. Tragically, there are many examples of this as well with some being fairly recent. Events, such as the 1998 U.S. Embassy bombings in East Africa, the U.S.S. Cole bombing in Yemen in 2000. Both World Trade Center bombings in New York in 1993 and 2001. The Pentagon bombing and the Shanksville Pennsylvania plane high jacking crash also in 2001.

And most recently, the U.S. Embassy attack in Benghazi, Libya in 2012. These are just some of the many attacks that have occurred worldwide and continue to this day.

As an aside, I'd like to illustrate that even some incidents, that would seem to be apparent terrorist acts sometimes are not always classified that way. For instance, the 2009 attack at Fort Hood Texas, by Major Nidal Malik Hasan. The major had been tracked by the military for some time as a result of his online contact with radical Muslim clerics. He had also made many comments about the U.S. Military killing Muslims.

When he started his attack he shouted "God is great" in Arabic then opened fire and killed 13 soldiers wounding 32 more. This incident was investigated by the F.B.I, the Department of Defense and the Senate Select Committee.

Although it was controversial at the time the Department of Defense classified this incident as a "Workplace Violence Episode". Nothing more. Major Hasan was convicted of murder and sentenced to death for his actions by a military court. He is currently awaiting his sentence to be carried out. (In 2015, the Federal Government, reclassified this incident as a terrorist act)

Another example that does not fit neatly into a box is the 2013 Boston Marathon Bombings. The suspects were identified by the FBI as two brothers from Chechnya. During an initial interrogation in the hospital, after his apprehension,

*When Disaster Strikes*

Dzhokhar, the younger brother alleged that his older brother Tamerlan was the mastermind. He said they had been originally motivated by extremist Islamist beliefs and the wars in Iraq and Afghanistan and that they were self-radicalized and unconnected to any outside terrorist groups.

This was later dismissed as secondary to the fact that Dzhokhor had been unable to assimilate like his older brother and was frustrated.

Was this Domestic Terrorism, Transnational Terrorism or simply criminal activity by two disillusioned immigrant brothers looking to lash out at something? Regardless of the label, innocent people were killed and maimed.

One final example that also does not fit neatly into a specific category is the 2015 attack in San Bernardino, California. On December 2, 2015, 14 people were killed and 22 were seriously injured in this attack which consisted of a mass shooting and an attempted bombing. The perpetrators, Syed Rizwan Farook and Tashfeen Malik, were a married couple. Farook was an American-born U.S. citizen of Pakistani descent and Malik was a Pakistani-born lawful permanent resident of the United States. Although called a terrorist attack by local authorities they have resisted in admitting any further involvement of outside groups or conspirators, even though their seems to some evidence of that.

On a personal note, my wife and I were on vacation and having dinner in Paris the night of the November 2015 terrorist attacks. Fortunately, we were on the other side of the city and not physically affected but still shaken up by the events that transpired. We were able to return home safely the next day.

The topic we need to discuss next has a direct correlation to terrorist acts. In essence, it is the tools or means of perpetrating a terrorist event. I'm speaking of *"WMD"*, weapons of mass destruction. In some emergency management circles the term WMD is referred to as "Weapons of Mass Death" or mass casualty.

The reason for this difference is that not all WMD's destroy property, but their main intent is always to take lives. For instance, chemical or biological attacks do little or no property damage.

According to F.B.I. statistics, firearms, explosives and readily available incendiary products are the most likely things to be used by terrorists. A case in point is the Boston Marathon bombing which used homemade pressure cooker bombs. Using mostly ordinary household items they built and detonated two bombs. Following directions they found on the internet, the two perpetrators were able to put together these devices, place them along the parade route and then detonate them, terrorizing an entire city.

Unfortunately there are many other ways for terrorists to deliver their intended goals of death and destruction. They have adapted their operations as well as their modes of weapon delivery. And whatever their avowed goal is, the results are generally the same; death and destruction of both their perceived enemies as well as innocent people. These people are driven and don't operate under the same principles as regular society.

To illustrate and list those ways, an acronym has been developed known as *"CBRNE"*.

*Chemical*

*Biological*

*Radiological*

*Nuclear*

*Explosive*

Chemical weapons include any of the following; Nerve agents (Sarin), Blister agents (Phosgene and Mustard gas), Blood agents (Cyanide), pulmonary agents (Ammonia, Chlorine), Irritants (oleoresin capsicum, tear gas) and Toxic industrial chemicals. Unfortunately, Chemical weapons have been used by governments during wartime. We have seen them used by both sides during World War I (mustard gas), Sadaam Hussein in Iraq against the Kurds (Sarin) and in Vietnam where the chemical nicknamed *"Agent Orange"* was used as a defoliant in the jungle. Unfortunately, it also caused adverse health issues for many of the American armed forces who were deployed there. And now we have seen chemical weapons used again in the Syrian Civil War. Obviously, I regret to say, there are many, many more examples in each of these groups.

Healthcare facilities ability to respond and treat victims of Chemical weapons exposure is somewhat limited initially, until the substance can be identified. How do you protect staff, how do you prevent further facility contamination? Once identified, a proper course of treatment and dispensation of medication or antidote can be prescribed. The bigger issue sometimes in this type of exposure, is not the number of actual exposure victims, but rather the ancillary ones.

The *"worried well"* (individuals not truly affected, but who respond to the hospital nonetheless) can many times have a far greater impact on a hospital or healthcare centers' ability to operate simply due to the large patient surge that can occur. The facility then needs to sort through everyone presenting and determine who is in actual need of immediate care and who can wait. And while all of this is going on, the facility is still trying to operate for the rest of its patient population.

The panic and fear that this type of incident generates should never be underestimated. A good documented example of this is the 1995 Tokyo Subway attack. In five coordinated attacks, the perpetrators released "Sarin" on several different lines of the Tokyo subway system.

They killed 13 people, severely injured 50 and caused temporary vision problems for nearly 1,000 others. On the day of the attack, ambulances transported 688 patients and nearly five thousand people reached hospitals by other means. Hospitals saw 5,510 patients, seventeen of whom were deemed critical, thirty-seven severe and 984 moderately ill with vision problems.

Most of those reporting to hospitals were the "worried well", who had to be distinguished from those who were actually suffering from the effects of the Sarin. As can be expected, emergency services, including police, fire and ambulance services, came under criticism for their handling of the attack and the injured. Health services in particular, including hospitals and health care staff were also criticized: one hospital refused to admit a victim for almost an hour, and many hospitals turned victims away.

They later stated that they felt their services were not adequate to treat victims with this nature of injuries. In reality, it was later surmised that the hospitals' staff were simply overwhelmed with fear, including the physicians and administration.

*St. Luke's International Hospital*, which was located in Tsukiji, was one of a very few hospitals in Tokyo at the time that was set up for conversion into a "Triage Center" in the event of a major disaster. Currently only Japan and Israel have these formal capabilities.

This was indeed fortunate as the hospital was able to receive and treat over six hundred victims from Tsukiji Station, which was one of the locations of the attack. As a result of St. Luke's response, no fatalities were attributed to victims from that station.

This example in a nutshell, illustrates both the positive and negative response of healthcare facilities to a given disaster. The next group of WMD are *Biological Weapons*. These are developed from living organisms and can be weaponized to maim or kill. They can and have been utilized against humans, animals and even agriculture.

Biological warfare is not a new concept. It has its origins in *antiquity*; Persian, Greek, and Roman literature from 300 BC quotes examples of dead animals used to contaminate wells and other sources of water. In *Medieval times,* it was reported that bodies of victims that had died from *"the plague"* were dumped into water sources or thrown over town wall's that were under siege in the hope of killing even more people. Biological agents were developed and used in the French and Indian Wars, World War I and World War II.

"Germ warfare" as it was known in the 1950's was developed as a strategic weapon during the "Cold War" by governments on both sides of the "Iron Curtain". The bigger threat today, is not so much the danger of governments using biological agents, but rather them falling into the hands of terrorist groups around the world. One of the problems, with biological warfare is that the weaponization of a biological agent is not that easy. Its basic characteristics make it less than the ideal weapon agent. In addition, once dispersed, the biological agent cannot always be controlled as to its effectiveness, individuals affected or its spread pattern.

This at least is some small consolation for the general public at large. But we must always be aware of the possibility, and continue to prepare and protect against the danger. At this time we'll briefly discuss, *radiological weapons.* A radiological weapon or radiological dispersion device (RDD) is any weapon that is designed to spread radiological material. It is done so with the specific intent to kill and cause disruption. Also referred to as "dirty bombs", these devices set off by conventional explosives are then intended to disperse radioactive material in area to maim or kill as many people as possible.

While not necessarily an intended target of such devices, hospitals and healthcare facilities do contain many of the materials that could be used to make one. As a Healthcare Emergency Manager, you need to be attuned to this potential facility risk. You need to work hand in hand with the Coordinator of Radiology, the Facilities Manager, Director of Environmental Services and the Director of Security to identify, safeguard, secure and dispose of any items that the facility has that could be used in this type of device.

In particular, what is your facility's policies on receiving, storing, using and disposing of radioisotopes? This pertains to not only the isotope itself but to the materials used in a patient room for treatment as well. These items may also become "hot" as a result of exposure. Does your facility use radioisotopes for research work? What becomes of the product when the research is completed? Do you store it on site? If you do, how do you secure it? If not, what is the disposal procedure for them? These are the questions you need to ask and answer as a Healthcare Emergency Manager to protect your facility, staff and patients.

I'd like to recount a quick story here to illustrate just how important it is to track, handle and properly dispose of all exposed "hot" waste material. A patient had been treated at my facility using a radioisotope.

The treatment area had been lined with "chucks" (disposable absorbent pads). After treatment, the "hot" room was cleaned according to protocol and all material disposed of in the proper manner. Or so we thought.

We compacted our general waste in our receiving tunnel area and it was subsequently picked up by a carting company and brought to a waste site. This particular time, the truck was stopped at the entrance to the waste facility for having set off their radioactive detector alarm. The truck returned with its load to our facility. The compactor's contents were then "dumped" into a back parking lot on our campus.

The items were then sifted through using radiation detectors. In the middle of the load, a single disposable "hot chuck" was found. It was then disposed of properly. Environmental service staff then had to clean up all of the remaining waste and dispose of it properly as well. The time, use of staff, and subsequent fine levied against the facility was an expensive and needless lesson to have learned had the original cleanup been handled properly.

The next category in CBRNE is *Nuclear.* While in and of itself, the nuclear threat is not a top priority of Healthcare Facilities for preparedness, facilities do need to be aware of the risk should a nuclear event occur?

Obviously, if an event does occur and the Healthcare Facility is still functioning, its main focus would be the treatment of victims. Does your facility have a policy for dealing with victims of fallout and radiation sickness? What is your isolation area capacity? This is where you will need to consult with Radiology staff as well as your Emergency Department Chairperson and Trauma Coordinator. Also make sure that the Nursing Department is briefed into all of your planning.

Nurses are the first line of defense and are the people in the trenches. Maybe most important of all, the question you need to answer is how do you protect staff? Always remember, staff are our biggest asset. Protect them at all times, because without them we can't perform our mission for ourselves or our community.

The final category we will discuss here is *Explosives*. I'm fairly confident that almost everyone serving in the capacity as a Healthcare Emergency Manager is aware of the potential damage and destruction that could result from an explosive device being detonated within their facility.

Depending on the size and placement of the device, it could wreak havoc on a facility and potentially cause the deaths of numerous individuals. Suffice it to say, it's something none of us would ever want to have to deal with.

But something we do have to deal with is the *"bomb threat"*. This incident by itself can be extremely stressful and disrupting for the staff and facility. It must always be taken seriously, Law Enforcement should always be appraised and involved and staff need to receive proper training prior to any incident.

You need to have a good policy in place, to direct staff as to what their role is, regarding a bomb threat, without putting them at unnecessary risk. In all my years in Healthcare Public Safety, Security and Emergency Management, and after close to 30 bomb threats, I am happy to say none of those turned out to be founded.

 In fact, in researching this book, I could find no documented case of an explosive device being intentionally detonated within a hospital. That, at least, is some good news, although we need to remain on guard and vigilant at all times.

On the other hand, there was a recent incident involving an explosion (unintentional) at a Maternity Hospital in Mexico City. A propane truck that was refueling several gas tanks at the hospital exploded, destroying almost twenty-five per cent of the facility. Unfortunately, this disaster also resulted in deaths and injuries of both staff and infants. This was a terrible occurrence that hopefully, none of us will ever encounter. But again, you must prepare for almost any inevitability.

As this time, I'd like to share another personal incident that I believe is relevant to what's being discussed here. Approximately twenty years ago, while I was working in the Security Department at my facility. It was a Saturday 8-4 shift and had been fairly quiet. A housekeeping employee came into the office and handed the supervisor at the desk a cylindrical object. We both laughingly stated, "Oh, it looks like a pipe bomb". As a matter of fact, that's exactly what it turned out to be. The employee had found it behind one of the radiators in an old outer building. He then walked it through the connecting tunnels and into the main tower building, going through the Emergency Department and then to Security.

It was subsequently turned over to the Police who in turn had their "Bomb Squad" pick it up. The Police took possession of the device and it was later detonated by the Police at their range. This incident was never called in as a bomb threat, the device was found purely by chance, and by the grace of God it never detonated anywhere within our facility. It was pure luck, which as an Emergency Manager, is not something you want to depend on. While it's impossible to list every known hazard, I have attempted to list the more probable ones that you may have to face. If you haven't already, now is the time I hope you're asking yourself questions about your own facility and relevant preparedness level.

At this point, having discussed some of the multitude of hazards we currently face in healthcare, I would like to take the time to speak about the victims and injuries we may be treating. When evaluating your own facility's capabilities, you must obviously look at facility size, bed capacity, surge capacity, Emergency Department capacity, treatment level (trauma level) and overall services provided. Know what you're capable of handling before an incident occurs.

Questions to ask yourself; is your facility equipped to handle a bio-hazard incident? Remember about your ability and responsibility to protect your staff. Do you have the proper personal protective equipment (PPE) on hand for them? Have they been trained in its use? Can you properly treat burn (chemical and fire related) patients. Do they need to be transferred? What about major trauma, what level is your facility? How about mass casualty events?

Five victims for some facilities might create a surge scenario, for others it may be twenty. Have you explored setting up additional triage areas or treatment areas within the facility?

Are they on or off site? Can they be activated quickly? Are they stockpiled? What is their proximity to your Emergency Department? And what about blast victims? You'll have trauma, fractures, burns and concussive injuries. Let's take a minute to talk about concussive injuries.

Victims in the close proximity of a blast are faced with the additional hazard of overpressure (shock wave) from the explosion.

In general, primary blast injuries are characterized by the absence of external injuries; thus internal injuries are frequently unrecognized and their severity underestimated. They may appear to have no injuries other than their hearing being affected (eardrums are the first things damaged by overpressure). But in reality, these victims may have suffered some of the most severe injuries.

The lungs, and the hollow organs of the gastrointestinal tract are also extremely susceptible to these type of injuries. Gastrointestinal injuries may present after a delay of hours or even days after the initial event. One of the most severe injuries that can occur is *Blast Lung.* The term *Blast lung* refers to severe pulmonary contusion bleeding or swelling with damage to alveoli and blood vessels, or a combination of both of these. It is the most common cause of death among people who initially survive an explosion. This listing of injuries is an overview on what you are most likely to be dealing with as the result of an explosion.

When you evaluate the hazards your facility may potentially face, using a *Hazard Vulnerability Analysis (HVA)* or other tool, will allow you to identify your facility's capabilities, strengths and weaknesses. Sit down and take the time to think about what you might have to face at your facility.

Besides the obvious weather issues for your region, what else is out there posing a potential risk? What have other facilities in your area dealt with? What is manufactured and or stored in your community? Are you near a transportation hub; roadways, trains, shipping?

What is the history of those type of events in your area? An important first step in this process is to reach out, get to know and network with other Emergency Management professionals in your area. The last thing you want to be doing is to be introducing yourself to someone during an incident and then asking for help.

Start with other area hospitals and healthcare facilities, your local Police, EMS and Fire Departments. The local Building Inspector, Fire Marshal and Health Department. And don't forget about professional associations, trade publications and of course, the internet.

As we move forward in this process, you'll learn that contacts will be one of the most important resources to have in your tool box. We'll now move on to the question as to **why** we need to prepare.

## CHAPTER 2

## Why are we preparing?

I guess the simple and easy answer would be to say, because we have too. While that is partially true, the real reason is not that simple or easy to state. For the Healthcare Emergency Manager it's so much more than that. It's one thing to "meet" standards for compliance for your facility, it's another thing to fully understand those standards, what they represent and the ramifications for not meeting them. Not just in citations or fines, but in the real cost of property and lives.

Being an Emergency Manager in any field, but especially in Healthcare is more about following a calling, almost like a vocation, as opposed to simply being a professional in that field of endeavor. As such, the commitment level and sense of responsibility tends to be higher. In healthcare, we have two separate communities to watch over and protect. Our external community, the people of the towns and cities we serve and our internal community. These of course, are our patients and staff. You as the practitioner, need to make preparedness a priority for yourself and your facility.

As we go further into this chapter, we'll speak about the vast amount of regulatory agencies we must contend with; from Federal regulations and guidelines, to industry standards, all the way down to the local Health Department.

For now, just understand, that not only what we do, but more importantly, how we do it, will to some extent determine the outcome of our response to a disaster.

To that end, one of the most important things we can do to prepare is to train and educate not only our line staff, but supervisors and managers as well.

And let's never forget about Administration. We need to educate and involve them in every step of the process. We also need their buy in and support as well as a personal comfort level for them in the roles they will play during any disaster response. Make them not only allies, but proponents of your efforts as well.

That is one of the secrets of true preparedness. Not just meeting regulations, not just having your plan and manual in place. Not just conducting your drills and exercises. But more importantly, it's having staff that know what to do and when to do it. That will make all the difference in an actual event.

Healthcare facilities, and in particular hospitals that receive funding (billing & reimbursement) from the "Centers of Medicare and Medicaid Services" (CMS) must be *"accredited"*. Currently, the Centers for Medicare and Medicaid Services (CMS) has given deeming authority to four hospital accrediting organizations: *The Joint Commission (TJC); the Healthcare Facilities Accreditation Program (HFAP), the Det Norske Veritas Healthcare, Inc. (DNV) and Center for Improvement in Healthcare Quality (CIHQ).*

We'll start first with the most widely known of the four, *The Joint Commission (TJC).* Founded in 1951, as the Joint Commission on Accreditation of Hospitals (JCAH) its main focus was to aid hospitals in providing better patient services. Having proven itself, The Joint Commission was the first to be granted deeming authority by CMS in 1965. In 1987 the organization changed its name to the "Joint Commission on Accreditation of Healthcare Organizations" (JCAHO).

This was an effort to reflect its changing role and its expansion of services in the Healthcare industry. Then in 2007, again to stay current and to update their role in the industry, they adopted a new corporate name and logo. It is now known simply as "The Joint Commission" (TJC). The Joint Commission developed the tracer methodology to follow and evaluate the quality of a patient's healthcare experience. It is currently the largest accrediting organization with over 20,000 hospitals, healthcare facilities and programs surveyed across the country. While TJC unannounced surveys come approximately every three years, suffice it to say, we should always be in compliance with the regulations. This is especially true with the Emergency Management Standards (EM). The Joint Commission Emergency Management Standards have evolved as well and continue to evolve to this day. Initially, in their most basic form these began as part of the old "Plant Technology & Safety Management" (PTSM) standards under the "JCAH". Now, in their present form, the EM standards are tied to a larger group of standards known as "Environment of Care" (EC). After hurricane *"Katrina"* hit land in 2005 and devastated the gulf area, TJC reevaluated almost all of their standards. As a direct result, the EM standards increased dramatically in both number and detail. Now, almost every year, there are changes to these standards. For the year 2014 alone, there were twenty-five (25) proposed standard changes. The EM standards, like all Joint Commission standards are not meant to be static, but rather an on-going process that evolves given societal changes.

Now let's talk a little about *"Healthcare Facilities Accreditation Program" (HFAP)*. HFAP was conceived in 1943 and began surveying hospitals in 1945.

Initially, HFAP provided Osteopathic Hospitals with an opportunity to be accredited by CMS. This would ensure that osteopathic residents received training in facilities providing high quality patient care. In the mid-1960s, the United States Congress decided that accredited hospitals would be deemed to meet the conditions of participation.

They could then automatically participate in the newly established Medicare and Medicaid programs (CMS). HFAP quickly applied for and was granted deeming status in 1965. The HFAP standards are directly tied to the CMS Conditions of Participation (CoPs). Their accreditation process which is similar to TJC, involves a review of all HFAP standards during an unannounced survey every three years. Lastly, since the year 2014, also like TJC, a Life Safety (NFPA 101) Expert will be part of the survey process.

One big difference however, is in HFAP's survey process, which involves a comprehensive, non-biased and thorough review of patient-centered processes within the facility which are conducted in the least disruptive way possible. Another major difference, is the number of facilities accredited by the two organizations. With approximately 250 facilities surveyed, HFAP is dwarfed by TJC.

The third organization that has been granted deeming authority by the Federal government is *"Det Norske Veritas"* (DNV) (The Norwegian Truth). DNV was originally organized as a foundation, with the objective of *"Safeguarding life, property, and the environment"*. The organization's history goes back to 1864.

This was the year when the foundation was established in Norway to inspect and evaluate the technical condition of Norwegian merchant vessels. DNV provides services to many industries including healthcare. DNV's accreditation program is called the National Integrated Accreditation for Healthcare Organizations (NIAHO). The DNV process also integrates CMS CoPs and compliance with the International Organization for Standardization's ISO 9001. That is one key difference between TJC and DNV. ISO 9001 deals with the requirements that organizations wishing to meet the standard must fulfill. DNV has a long history in ISO certification in health care by some overseas health care systems.

 Another key difference between DNV and the Joint Commission is that DNV surveys its member hospitals annually and not every 18 months to three years like TJC. DNV's survey approach is based on a cooperative effort with the facility as a partner, rather than a punitive approach. Their focus is to find ways to perform better as a facility as opposed to finding what's wrong. DNV received CMS deeming authority for hospitals in 2008, making it the third accrediting option for hospitals.

And like HFAP, DNV with about 500 hospitals surveyed is also overshadowed by the TJC. There is an excellent article online showing the differences of the first three organizations titled; *"The Big Three" a side by side matrix comparing Hospital Accrediting Agencies, by Diane Meldi, Faith Rhoades, and Annette Gippe.*

The final Organization having deeming authority is the *"Center for Improvement in Healthcare Quality" (CIHQ).* Established in 1999, the Center for Improvement in Healthcare Quality (CIHQ) is a membership-based organization comprised primarily of acute care and critical access hospitals.

They are headquartered in Round Rock, TX. CIHQ was granted deeming authority in July of 2013 for an initial four year period. CIHQ are currently authorized at this time only for "Acute Care Hospitals", not psychiatric or critical access hospitals.

Regardless of the accrediting organization your facility utilizes, it's important to remember that each facility is different, so some of the preparation process may actually turn out to be trial and error for you. Don't be afraid to be creative in your approach, and by all means make sure you involve all appropriate staff throughout the preparation process. Don't ever try to go it alone. There is no magic formula that works the same for everyone. Nor should there be. So good luck.

Over my almost forty year career in the healthcare industry, the facilities where I have been employed, were all surveyed by The Joint Commission. So that is the organization that I have the greatest knowledge and understanding of. While I wouldn't presume to try and tell anyone how or what to do to prepare for a Joint Commission or any other survey, I would like to relate some of the things that made the process easier and more effective for myself, my facility and the surveyor. It is my sincere hope that you find some of this information useful to you and your facility. If it is, please feel free to use it.

First and foremost in my opinion, is that the facility should have a specific individual handling Emergency Management. With the expansion of TJC Emergency Management standards, gone are the days, especially in a large facility, where this responsibility could simply be an assigned duty to someone else, such as the Security Director or the Safety Director. Ideally, the individual should be knowledgeable in the field and preferably be credentialed. That way your facility is protected by an industry professional.

And while TJC does not specifically require it, it lets TJC surveyor know that the facility takes its role as a community asset seriously by having a separate Emergency Management Professional. It's been my experience in the survey process that the TJC surveyor usually asks who the Emergency Manager is and not who covers emergency management. That's because they want to know who they'll be dealing with when they run an exercise during the survey. Having stated that, if this is not feasible, then a conscious effort should be made to find the most qualified individual within your facility to fulfill this role. Hopefully, it will also be someone who is actually interested in preforming this duty. It should never be "dumped" on anyone. This does a disservice to both the individual and the facility.

Secondly, let's discuss the *"Emergency Management Committee"* itself. Do you have one at your facility? Who sits on the committee? How often do you meet? These may seem like basic questions, but believe me, it's been my experience that much more than expected, facilities don't have a separate Emergency Management Committee. These facilities run Emergency Management out of the *"Safety Committee"* or out of an *"Environment of Care Committee"* (EC). One smaller facility I visited even covered their Emergency Management issues through their *"Workplace Violence Committee"* (WPV).

While you need to set up the format that works best for you and your facility, I would strongly suggest that you maintain a separate and distinct facility Emergency Management Committee. And while the Emergency Management committee should work closely with Safety, EC and WPV, the facility is best served when it operates as an independent entity.

Although TJC no longer mandates it, TJC has set up a separate set of standards for EM, so this is generally something they usually like to see. It shows the surveyor that the facility is conscious of and takes seriously, the myriad EM issues that face any Healthcare Facility today.

Now I would like to address the make-up of the committee itself. Again, TJC no longer specifies who needs to be on the committee. They do however, make references regarding staff participation. Specifically, that the senior staff must be part of and involved in both the planning process and review process. This then, allows each facility some leeway and options as to who to place on their committee. There should be some basic ones included; Facilities, Security, Environmental Health/Safety, Medical staff, Administration, Nursing, Communications (IT) and Emergency Department. There are some others to be considered in this mix; Infection Control, Pharmacy, Laundry, Food & Nutrition. Whatever you decide, just find the combination that works best for your facility. This will be a little bit of a trial and error process.

Try to have people assigned to committee who are actually interested in being involved with the process and not just sent there to fill a chair or position. Also, don't be afraid to rotate members on the committee, its keeps it fresh by presenting new eyes and ideas. Consider having some individuals just give quarterly reports to the committee. This makes participation less burdensome for them.

The next suggestion pertains to the frequency of the EM committee meetings. Assuming you have your committee set up, the next thing you want to address is the frequency of your meetings. First of all, the committee should meet the same time and or day for each scheduled meeting. For example; the second Wednesday of the month at 9:00 AM. Select a routine.

This will make it easier for the members to schedule and attend the committee meetings. What worked best for my facility was monthly meetings. If your facility is relatively small, say less than one hundred (100) beds, than you could probably go with bi-monthly meetings. I would not suggest stretching them out any further than that.

Finally, regarding the Emergency Management Committee, several things of importance should be kept in mind; have a set agenda, take good notes, keep/publish/distribute meeting minutes, take/keep track of attendance and finally make sure you follow up on issues identified at previous meetings. The committee needs to resolve or closeout these issues. Do not leave them unresolved or lingering from month to month. This in particular, is something TJC surveyor will focus on and call you to task for.

Now I'd like to make a few suggestions about documentation. Nothing makes a Joint Commission surveyor happier than thorough documentation. Make sure that it's complete, well organized, and easy to reference. Always make sure that you are well versed in its content, set-up and presentation. Rehearse your material if you can, with someone from the Emergency Management Committee or with someone from your facility Accreditation Preparation Team. Be confident in your own ability when dealing with the surveyor. One of the worst things you can do is to be fumbling around trying to find a particular bit of information, saying I know I have it, while the surveyor looks on. I realize that this is a stressful time, even for a seasoned Emergency Manager. So, if possible, have your documentation set up both electronically and with hard copies. Keep the electronic version on your laptop and on a separate flash drive. As for the hard copy documentation, it takes a little more effort, but have everything in a *tabbed* binder. Have it set up by standard.

Cite the standard, answer whether or not you are compliant, then list where or what document verifies this, and include a copy of that documentation. When the surveyor asks about a specific standard, bring it up on the computer and go right to the tab in the binder. Believe me, this will not only impress them with your knowledge and efficiency, but will make it easier for them. If it's easier for them, they'll be happier, and if they're happier in the process you will be too. I sincerely hope that some of these suggestions make the entire survey process a little easier for you to deal with and get through. A good thing to remember is, no one has all of the answers, so again, find what works best for you and your facility.

Now let's move on to some of the Federal standards, regulations and guidelines we must deal with. Let me preface this by saying, that many of these guidelines while not an intimate part of our daily lives as practitioners are nevertheless important. At the very least, we should have both an understanding and knowledge of them, and what they represent. Having stated that, let's just jump right on in and begin with *"The National Incident Management System"*, more commonly known to us as NIMS. Originally issued in March of 2004 and revised in 2008 by Homeland Security Presidential Directive 5 (HSPD-5), "NIMS integrates existing best practices into a consistent, nationwide, systematic approach to incident management. This is applicable at all levels of government, nongovernmental organizations, and the private sector, and across functional disciplines. It is to be utilized in an all hazards context, regardless of cause, size, location of incident, or complexity of incident. There are five major components that make up this systems approach: *Preparedness, Communications and Information Management, Resource Management, Command and Management, and Ongoing Management and Maintenance.*

What NIMS is not, is an operational incident management plan, resource allocation plan or communications plan. The flexibility of the NIMS components allows you, as the user, to adapt to almost any situation, large or small, real life or planned exercise. There are two main items that are directly applicable to healthcare; *credentialing and training.* First credentialing; NIMS has developed credentialing requirements for forty-four (44) medical and public health functions. This allows facilities in their time of need to utilize other healthcare providers who have been pre-registered and certified. This came about as a direct result of issues encountered (particularly in New York City) during the September 11, 2001 terrorist attacks.

The other item is *training.* Specifically, the NIMS Incident Command System (ICS) courses. And while there are many ICS and NIMS courses available, the basic ones needed for you and your staff are; *ICS 100 Introduction to Incident Command, ICS 200 Single Resources and Initial Action Incidents, ICS 700 National Incident Management System (NIMS) an introduction, and ICS 800 National Response Framework (NRF).*

Rather than get into a discussion about each of the courses separately, or explain them in detail, you can find a complete list of all of these courses as well as others along with their course description online.  They are on the *National Emergency Management Institute* website. As a matter of fact, you can take all four of these courses and many others online through the EMI website.

Now let's speak a little bit about the *National Response Framework* (NRF). Originally begun as the National Response Plan, the National Response Framework was issued in January of 2008. The *National Response Framework (NRF)* is a *guide* to how the Nation conducts an *"All Hazards"* response.

It is built upon scalable, flexible, and adaptable coordinating templates to allow for the alignment of key roles and responsibilities across the Nation, linking all levels of government, nongovernmental organizations, and the private sector. It is intended to capture specific authorities and best practices for managing incidents that range from the serious but purely local, to large scale terrorist attacks or catastrophic natural disasters. For more detailed information, please download the entire document from the web.

In an effort to define what National Preparedness actually means, Presidential Policy Directive * (PPD-8) was established. This originally came out of Homeland Security Presidential Directive eight (HSPD-8) under President George W. Bush. These guidelines, have now been updated amended and superseded by *Presidential Policy Directive 8,* signed by President Barack Obama on March 30, 2011. Its purpose and intent is to fortify national security and the resilience of the United States of America during times of disaster or national crisis. This is accomplished in the following way, national preparedness becomes the shared responsibility of all levels of government, the private and nonprofit sectors, and individual citizens. Everyone can contribute to safeguarding the Nation from harm.

As such, while this directive is intended to galvanize action by the Federal Government, it is also aimed at facilitating an integrated, all-of-Nation, capabilities-based approach to preparedness" (Taken directly from PPD-8). The focal point for PPD-8 now is the requirement for all parties to be involved in the process, a national community approach and not just the Federal, State or Local Governments. In addition, it utilizes a risk-based regarding preparedness support. Next, it builds core capabilities to address all disasters.

This is accomplished by working with the entire community to build these core capabilities to counter any crises that may be faced, those communities become full and working partners with government, as opposed to simply being bystanders waiting to be told what to do. It also integrates those efforts over all phases of emergency response; prevention, protection, mitigation, response and finally recovery. And finally it requires assessment of performance outcomes in order to gauge our effectiveness.

PPD-8 is then broken down into the following parts;
*"The National Preparedness Goal"*
*"The National Preparedness System"*
*"The National Planning Framework"*
*"The National Preparedness Report"* (compiled annually)
*"The National Campaign to Build & Maintain Preparedness"*.
While I'm not going to go into detail for every one of the above, I would like to take a quick look at the "National Preparedness Goal" (NPG). The NPG actually describes what it means for the entire or whole community to be prepared for any and all disasters. The NGP is very specific; *"A secure and resilient nation with the capabilities required across the whole community to prevent, protect against, respond to, and recover from the threats and hazards that the greatest risk"* (DHS 2011).

There is another Presidential Directive I want to brief you on, and that is *"Presidential Policy Directive-21"* (PPD-21). PPD-21 is the Country's national policy for the enhancement of protection for the nation's *"Critical Infrastructure"* (CI). A key component of PPD-21 is the *"National Infrastructure Protection Plan "(NIIP)*. Basically when you think of infrastructure you think of; water, power, communications, transportation, public health and safety. Everything that affects our daily lives, and then some.

This has already evolved into the *"The National Plan"* which is simply an update of the NIPP. It builds on the previous NIPP and reflects the many changes in critical infrastructure risk. It integrates cyber, physical and human elements in CI to manage risk, with an increased emphasis on cyber.

Another set of Federal requirements that must be complied with are the ones from The United States Department of Labor. I'm referring of course to the Occupational Health and Safety Administration (OSHA) regulations.

Specifically related to healthcare and encompassing emergency management are the Code of Federal Regulations, *29 CFR 1910.1200*, which is the *Hazard Communication Standard*. This standard is designed to ensure that employers and employees know about hazardous chemicals in the workplace and how to protect themselves. Employers with employees who may be exposed to hazardous chemicals in the workplace must prepare and implement a written Hazard Communication Program and comply with other requirements of the standard.

The next is *29 CFR 1910. 1030* the *Bloodborne Pathogens Standard*. OSHA issued this standard to protect employees from the health hazards of exposure to bloodborne pathogens. Employers are subject to OSHA's Bloodborne Pathogens standard if they have employees whose jobs put them at reasonable risk of coming into contact with blood or other potentially infectious materials. This can be especially true during disasters. Employers subject to this standard must develop a written exposure control plan, provide training to exposed employees, and comply with other requirements of the standard.

The third is *29 CFR 1910.38* the *Emergency Action Plan Standard.* OSHA recommends that all employers have an Emergency Action Plan. A plan is mandatory when required by an OSHA standard. An Emergency Action Plan describes the actions employees should take to ensure their safety in a fire or other emergency situation, including evacuations. Although these regulations are the ones that would most likely come into play during a disaster there are many other OSHA regulations that have a direct bearing on healthcare facilities.

The fourth one that I would like to mention is *29 CFR 1910.134 Respiratory Protection Plan.* In any workplace where respirators are necessary to protect the health of the employee or whenever respirators are required by the employer, the employer shall establish and implement a written respiratory protection program with worksite-specific procedures. There are many applications of this standard for healthcare workers. This will also certainly be the case during various disaster scenarios. The program must be administered by a suitably trained program administrator. The program shall be updated as necessary to reflect those changes in workplace conditions that affect respirator use.

Please take some time to review and familiarize yourself with all of these standards. It will be well worth your efforts. This abbreviated listing gives you some idea as to the many federal regulations that you must become educated and compliant with.

As far as State, City, County, Town and local Health Department regulations, it's understood that each jurisdiction is unique and has its own specific requirements and regulations. These are tailored to the individual communities they represent.

To be effective as a Healthcare Emergency Manager you must familiarize yourself with all of the jurisdictional regulations that will also impact your facility, and coordinate a plan for compliance. For assistance look to others in the field. To reiterate an earlier point I made, learn to network with your professional peers and build your base of contacts. They will prove invaluable to you in your time of need.

At this point we're going to shift gears and look at some of the industry standards that have a direct bearing on hospitals and healthcare facilities. We've already spoken about the accrediting agencies and why we need to be in compliance with them. We've also discussed some of the federal regulations and guidelines that we must contend with. Now we'll explore some Industry standards and guidelines.

What I'd like to discuss now, briefly is the *National Fire Protection Association* (NFPA). The NFPA which is based in Quincy Massachusetts, is not a government agency. Nor does it produce regulations or levy fines. Rather, it is a trade association that creates and maintains private, copyrighted, standards and codes for usage and adoption by local governments. The NFPA is responsible for some 300 codes and standards that are designed to minimize the risk and effects of fire by establishing criteria for building, processing, design, service, and installation primarily in the United States. They are nationally recognized and referenced, and are generally adopted by local jurisdictions.

They cover many different industries, including healthcare facilities, and the many different aspects of those industries. But for our purposes, we'll be focusing on healthcare only at this point and time.

The specific standards that I would like to mention here that have a direct bearing on healthcare facilities are; *NFPA 99 (Healthcare Facilities), NFPA 101 (Life Safety Code) and NFPA 1600 (Disaster/Emergency Management and Business Continuity)*.

*NFPA 99* establishes criteria for levels of health care services or systems based on risk to the patients, staff, or visitors in health care facilities to minimize the hazards of fire, explosion, and electricity. This standard also requires facilities to have an "Emergency Management Committee".

*NFPA 101* "The Life Safety Code" is the most widely used source for strategies to protect people based on building construction, protection, and occupancy features that minimize the effects of fire and related hazards. Unique in the field, it is the only document that covers life safety in both new and existing structures. In Healthcare, this is probably the most widely used of any NFPA referenced standard. In particular this is the one utilized by all Healthcare Accrediting Organizations and other local and state agencies when conducting surveys.

*NFPA 1600* originated from the "National Commission on Terrorist Attacks upon the United States (the 9/11 Commission)". It recognized NFPA 1600 as our National Preparedness Standard. It is widely used by public, not-for-profit, nongovernmental, and private entities on a local, regional, national, international and global basis.

NFPA 1600 has been adopted by the U.S. Department of Homeland Security. This is a voluntary consensus standard for emergency preparedness. Its provisions cover the following items.

They are development, implementation, assessment, and maintenance of programs for prevention, mitigation, preparedness, response, continuity, and recovery pertaining to emergency management. Like NFPA 99, NFPA 1600 also requires an "Emergency Management Committee".

If this is the first time you're hearing about the NFPA or these sets of standards, you my friend, have a lot of homework to do. The positive thing about it, is that if you wish, you can become a member of the NFPA. You can also get up to speed by attending any one of their informative seminars or by purchasing their codes and supplemental handbooks. As I mentioned above, the NFPA is not a regulatory agency. You won't be fined by them for not complying with the codes. But there may be repercussions down the line for not adhering to them. I'll explain what I mean by that here. If their (the NFPA) codes have been adopted by your local governing body or by the "Authority Having Jurisdiction" (AHJ), which they probably have been, you will want to make sure that you're in compliance.

As a quick side note, the Joint Commission also references the NFPA codes. I would think the other accrediting agencies do as well. So, for your sake as well as your facility's, make sure you become knowledgeable and comfortable in maneuvering through these particular standards. It will end up making your life so much easier if you do. Since these are also considered Industry "Best Practices", if you are not following them, as I stated, it could potentially present a problem. If something adverse happens at your facility and there is subsequent litigation, you may be called upon to testify on behalf of your facility.

Do you really want to be sitting in the witness box trying to explain to the plaintiff's attorney why you chose not to follow a recognized, referenced, industry standard? I know I wouldn't.

As you can plainly see, even by this abbreviated listing, there are many regulations, standards and codes that we must be in compliance with. With knowledge and understanding comes strength. Not only do you, as the Emergency Manager, need to know them, you need to make sure that your staff are well informed and knowledgeable as well. It all takes time, but none of it is insurmountable. Even the biggest pile starts with that first shovelful. So get ready to dig in and get your hands dirty. This will help you to develop an understanding and confidence level that you are going to need.

The next chapter will present some ways to help you decide on how to prepare your facility for a disaster. Things such as who you need to contact and network with, what kind of resources do you need and what "tools" will you use to accomplish the task?

Some of the things we'll also discuss will be outside vendors, "Memos of Agreement" (MOAs) and "Memos of Understanding" (MOUs), keeping and maintaining an Emergency Management supply inventory and being compliant with "The Joint Commission's" (TJC) *"96 hour rule"*.

# CHAPTER 3

## How do we prepare?

How do you prepare your facility for a disaster? That's an extremely important question, and one in which you, as the Emergency Manager must seek the answer to. In this chapter, I will provide some suggestions that will make that task somewhat easier for you. My intent is to point you in the right direction so that you can explore which things will work best for you and your facility. What I'll present here are things that I've developed myself, or learned about from other Emergency Managers and Healthcare Professionals. Let's face it, we all learn from someone. That's the importance of networking. Plus, we need to continue that learning process throughout our professional career, so that we can remain viable and current.

First and foremost, let's start off with your own facility staff. What is the history of events in your area? What have you identified as potential hazards? About fifteen to twenty years ago, Florida had an extremely bad hurricane season. If I recall, I think they had something like five storms make landfall. Yet many hospitals and healthcare facilities fared well with their staffing levels. One of the reasons they did, was that as a result of previous history, they instituted measures to safeguard and assist their staff. They realized that if staff had less to worry about, they were more likely to come in, stay longer and work harder. Some of the things they instituted were daycare areas, senior care areas and yes, even pet care areas for their staff. You take care of your staff and they'll take care of the patients.

In reality, not every facility will be able to accomplish this, but it is well worth mentioning and looking into. Now let's look in the mirror and see how you are regarded and perceived by the people you work for and by the people you work with? Are you the kind of person that flaunts your knowledge and experience or are you the kind of individual that works well with others? I hope you're the latter, as that's the one that will ultimately be successful.

Do you know and get along with all of the various department heads? What about your boss? What about line staff? Do you provide any educational sessions and or training to your staff? Do these people see you as helpful, as a problem solver? Are you a team player and consensus builder, or do you always have to be right? Remember, it's easy to go around telling people what they have to do, or what they're doing wrong. It's much better to help your facility staff. After all, they are part of your team and should be considered resources and assets to you. You can still educate and train them, but perform it in a positive and respectful way. Always make sure that you work with people to achieve your ultimate goal of facility readiness. Give them a format or template to work with in order for them to put together a Department Emergency Management plan. Set out the parameters of what you need and give a specific date you need it completed by.

Make yourself available to staff to answer any questions. Then review it with them so that it does not conflict with the overall facility plan. If there are changes to be made, try to make suggestions, and if possible, always remember to thank them for their effort. You'll be amazed at how much more cooperation and respect you'll receive by being polite, professional and courteous instead of confrontational or demanding.

This my friend is a big part of your job. It's part of why the facility hired you. Make sure that **you** are an asset and resource to your staff and facility at all times.

Let's start with one of the basics, the facility *"Emergency Operations Center"* (EOC). Is there already a designated area or do you need to set one up? If there is a designated area, is it a "shared space" or is it dedicated as the EOC? I've had both of these options at my facility and believe me, a dedicated area is the way to go, if at all possible. It gives you so much more control and flexibility. It also makes your response a lot more efficient. Plus, you will have the advantage of setting up computers and communication lines as well as anything else you deem appropriate ahead of time to meet the task before you.

Originally, and for many years our executive board room, which was located on the first floor of our twenty story tower was utilized for the facility EOC. One issue with that was that the room faced south and the entire wall was made of glass windows. We were always concerned about losing them in a major storm. Another problem was that during drills, exercises and actual activations if the room was being used, we first had to clear it out. Then we would open the "Mobile Disaster Cabinet" that we kept secured in the room. We would have to match up all the phones to their proper jacks, bring out charts, set up the computer ( 1 PC, not a laptop) put out pads, pens, pencils, markers etc. This all took time and manpower, and as we've mentioned earlier, that's not always a commodity you have. That all changed when we were assigned our own room, internally located on the third floor of the tower. We installed eight (8) PC's, one laptop, twelve phones, one satellite phone, two dry boards, two flipcharts, a small refrigerator, and a television connected to CNN and the weather channel.

We also had a monitor and video feed from Public Safety which allowed us to view the two entrances to the tower and the Emergency Department's parking lot entrance. There were property schematic maps of the hospital property as well as our Long Term Care facility posted on the walls.

Also in the closet, were all of the position vests for HICS-ICS, twelve two-way radios with six additional batteries, two cases of water and six portable cots. This gave us our own self-contained, self-sufficient unit and proved highly effective. And as a bonus, just outside the EOC, were a men's and ladies restrooms. Believe me this made life much easier.

Just as with your main Emergency Management supply and equipment inventory, everything was checked in our EOC for location and operability on a weekly basis. The last thing we wanted were people showing up to respond and not being able to function due to lack of proper equipment or supplies. Having a dedicated area/room for your EOC is the optimum way to proceed.

Now let's take a look at your physical plant and property footprint. First of all, what kind of facility is it? Is it long term, acute care, specialty care, critical access, community based, teaching facility, or major trauma medical center? Are you a stand-alone facility or are you part of a system or network? As you can see, even by this short list, the world of healthcare is pretty wide ranging in scope and services offered. Are you looking at a single building or multiple ones? Are those outer buildings connected to the main building or are they stand alone? Is it a combination of both? Do you have buildings off-site as well? What about the property itself? How big is it? Are there internal roads and main road access points? I've just posed a lot of different questions to you.

In reality, I'm positive that as the facility Emergency Manager, you've already have an answer to most of them. A good starting point would be to have as a resource a property map or schematic of your facility and grounds. Contact the Facility Manager or Engineering Director for this. Have you ever surveyed your campus? Is your facility on a master-key system or is it something more of a hodgepodge. Do you know which doors are locked, which areas or buildings can secured, and which internal roadways can be closed? Contact the Safety Director, Facility Director or your Security Director as they may already have the information you need. If not, depending on whether or not you have any staff assigned to you, speak to them about assisting you with this task.

Usually, when you explain that the information will be helpful to them as well, they will generally seem to be more receptive to the idea. Next, you should speak to your Facilities Director, to start building your informational file on the facility. Ascertain things like, acreage, square footage, power feeds, water feeds, exterior lighting, how many emergency generators do you have, both building and portable, how much fuel is on hand and how long you can run them. Then you want to know what's powered by the generators. You need to find out exactly what areas in the facility are covered.

How does the staff determine which outlets are on emergency power when they need to switch over? Conduct a facility reassessment if necessary, and then speak to the Facility director about the feasibility of adding additional areas as needed (this is an issue that could be brought up and discussed at an Emergency Management committee meeting). And if you have outer buildings, both on and off site, find out if anything is on emergency generators there as well.

This becomes particularly important if you're planning on using any of these buildings as *alternate care sites*. We'll discuss this in more detail shortly. What about those all-important contacts and networking system we keep speaking about. Whether you're new to the field and profession, or you're a seasoned veteran in healthcare, probably one of the most valuable things you can have is contacts.

Knowing who to call, for what you need, when you need it, can sometimes be the difference in succeeding or failing. Build your base of contacts slowly. Be sure that they are people you can count on during an emergency. Make sure they know they can also depend on you. It needs to be a two way street when it comes to trust and reliability. Start within your own facility first, then venture outside to seek other expertise.

The local Fire Department, the local Sherriff or Police Chief or Precinct Commander, your peers in other healthcare facilities, the local Health Department, your local Emergency Manager, your State Emergency Management Representative, the Salvation Army and Red Cross are good places to start. And when you're putting together your resource list don't forget to include the large warehouse outlets like *COSTCO, SAM'S CLUB or BJ's*. This list could go on and on, but build on it from here. Also remember to explore professional organizations like the *"International Association of Healthcare Security and Safety"* (IAHSS) *"International Association of Emergency Managers"* *IAEM* or *ASIS International.*

I am a former two time President of the Long Island Regional Chapter of the IAHSS. I can't begin to tell you how many times I reached out to my peers, or they to me for assistance on various issues.

You are part of a close knit community, take full advantage of that in a positive way. It makes all of us stronger and more prepared. There are also numerous "Blog Sites" dealing with Emergency Management and disaster response online that may be of interest to and assist you. Do some exploring on your own to see what you come up with.

I'd like at this point, to take some time and discuss *communications*. What are your current capabilities, resources and what you need to do to prepare? The one area that I always received the most feedback on whether it was an exercise or real life event was that of communications. Invariably, the consensus of opinion was that there could have been better communications during the event. I think we've all been here with this one. And sadly enough, there is a lot of truth to it, no matter how hard we continue to try to improve them. There is no quick fix with this issue.

The best that you can do is to just keep trying to improve your specific capabilities, and equipment. Again, start by doing a full facility assessment. Speak to your Communications Director if you have one. They will probably have most of the information you'll need and it will certainly save you time and effort. If not, you will have some foot work to do.

The following are some of the things you really need to be aware of; what are your land line capabilities? Are they internal or external? If external are they fed from more than one source? Is there a redundant system in case of primary failure? Is there battery backup and for how long? Can your land line system (battery back-up) be tied into emergency power? Does your facility utilize cell phones for strategic staff? Does the carrier provide any back-up?

One of the things we had at our facility that worked as an extreme back-up was a *"satellite phone"*. In the event that both land lines and Cell system went down we still had the capability of communication with the outside world. It's an expense, but one that's well worth the cost outlay.

There are two different government phone system communications programs you that should be aware of and participate in. They are the Government Emergency Telecommunications Service (GETS) and the Wireless Priority Service (WPS).

Both are offered by the Department of Homeland Security (DHS) Office of Emergency Communications (OEC). GETS, was developed in response to a growing need for priority communications for select users. GETS enhances call completion for select wireline (landline) users when abnormal call volumes exist. This is especially true during disaster response. WPS, was developed to address the growing need in the nation for priority communications on the cellular networks.

 As healthcare facilities, you are part of the community critical infrastructure and as such, eligible at no cost to participate in these programs. I highly recommend that you do so.

For additional information, contact the Department of Homeland Security Priority Telecommunications Service Center at 866-627-2255 or 703-676- 2255, via email at; GETS@HQ.DHS.GOV , or visit www.dhs.gov/gets or WPS@HQ.DHS.GOV, or www.dhs.gov/wps.

Some other means of communication that you could explore would be pagers, many facilities still utilize them.

This might be a good low cost back-up system for you to explore. Another is two way radios or "walkie-talkie". Find out if the Security or Facilities department has or uses them. If not, look into purchasing them. Local-band short distance radios are not a high cost item, and in time of need could be distributed to pre-ordained vital areas of your facility. Obviously, prior to use you would need to train staff on their use. You would also need to conduct an operability survey to find out if those radios work throughout your facility or if there are any "dead spots" (Look into doing this as well with your local Fire Department). There are two more ways I can suggest to communicate in your facility during a disaster response. One is basic and low tech the other one is a little bit "outside the box" so to speak.

 First the low tech basic one; *"runners"*. Runners are simply individuals that can be assigned to a specific area and then will take information to other areas. This can be set up well in advance of any actual disaster response. It's also something that you can easily conduct a *"drill"* on. (We'll speak later about the difference between drill and exercise). During a response, have the pre-selected, pre-trained individuals report to a pre-determined location. Areas such as your "Emergency Operations Center" (EOC), main lobby, or central nursing office are good places to consider for a variety of reasons. Each allows a distribution point that can feed information throughout the facility in time of need.

The "outside the box" means of communication I was referring to is *"Short Wave, Amateur or HAM radio"*. Having this type of system could prove invaluable during a disaster. You should look into both taking your basic license course and purchasing a portable unit set up for the facility.

The equipment is not that expensive for the average system you would need for your facility. Something else to consider, perhaps by going through your Human Resources department, would be to find out if there are any employees already licensed as amateur radio operators. They could be extremely useful to you and the facility in both operation and set up.

Another way to get information is to contact local amateur radio groups, clubs and associations. Again, they may be able to advise you on courses, equipment and operation. They can be a tremendous resource to you. The Amateur Radio Emergency Service (ARES) and Radio Amateur Civilian Emergency Service (RACES) are two such groups. ARES is activated before, during and after an emergency.

Generally, ARES handles all emergency messages, including those between government emergency management officials. RACES, on the other hand, almost never starts before an emergency and is active only during the emergency and during the immediate aftermath to assist government.

RACES is normally shut down shortly after the emergency has cleared. Both of these groups are sponsored by the Amateur Radio Relay League (ARRL). Another group, the Radio Emergency Associated Communications Teams (REACT) are volunteers who are organized internationally and use CB radios to provide public service communications for travelers.

Once again, I'd like to illustrate the above point on how important and practical it is to have this capability. Our Vice President of Information Technology (IT) was a gentleman by the name of Ron Tomo. Not only was Ron's expertise in IT (computer/communications) invaluable to me in my role as Emergency Manager but I came to find out that he was actually a hidden gem.

As we worked together first as colleagues, then as friends I came to find out that among other things, Ron was both an expert and instructor for amateur radio, having been involved with it for over forty years. He also had military experience and was well versed in disaster response. He readily jumped in with both feet to support what I needed. He offered to give classes to staff, coordinate any licensed individuals and help purchase and set up the type of system that would be the most beneficial to our facility. I couldn't have asked for a better partner. He ended up becoming an integral part of my Emergency Management Program.

At one point, after having spoken together about an upcoming State Emergency Management Communications exercise we would be participating in, Ron volunteered his time, equipment and expertise. He and one of his associates set up two portable units. One at our facility and one at another facility approximately five miles away. During the State exercise we passed information back and forth. As a result, our facilities were cited by the State for having gone above and beyond in making the exercise a success. I was extremely fortunate to have someone of his caliber in my corner and I'm proud to say we are still friends today. I can only hope that you find your own hidden gems within your facility. But you do have to ask and look.

One last thing I'd like to speak about is something we had set up at our facility that also worked very well. It was an *alert information* phone number. We had a dedicated phone number that could be reached both internally and externally by staff. We could record messages, warnings or updates as needed. We used this extensively during several construction phases when certain campus roadways or building entrances were blocked.

We also utilized it for all emergency management exercises and actual responses. We instituted it as a result of an actual lockdown situation that we experienced. During an actual event, someone would be assigned to update the recorded message either every hour or sooner if the situation warranted it. I hope some of these suggestions have provided you with some of the possibilities you can explore for your own facility.

As I've taught around the country, people have opened my eyes to many different possibilities; such as a facility in Utah using snowmobiles to communicate with off-site buildings, or a facility in Kentucky where horses were used if their roads were washed out. Remember, whatever works best for you and your facility to have communications during a disaster is the right way.

Now that we've done our best to prepare and protect our facility in the event of a disaster, let's take a look at *"alternate care sites"*. If you haven't already done so, the first place to look for and identify these is within your own facility. These are the most ideal, as they are the closest to you and the ones you have the most control over. You've completed your facility survey already so you should have at least a general idea on what potential areas you might be able to utilize during a disaster. A good thing to keep in mind when looking at other areas is not to focus solely on patient use. Be open minded, areas identified may also be used for staff, staff families or even supply and equipment staging areas. You'll want to start with a checklist of capabilities once you've identified an area.

Electrical outlet capacity, water facilities, medical gas, sanitation facilities, lighting, number of doors and their dimensions and the ability to secure the area are good places to start. Sanitation is an extremely important concern.

Just because an area has a bathroom, it may only be used occasionally. Now, you may be putting twenty or thirty people into an area for an extended period of time. Can the sanitation system handle that increased amount of usage? The same holds true for electrical outlets. How many are there, what amperage are they, and are they on emergency power?

The size of door frames comes into play if you are looking to move patient beds or equipment. Will they fit through the existing door opening? Find out ahead of time, so you won't be surprised or embarrassed during a disaster. And if you're going to have people sleeping in an area, make sure it can be secured for their safety.

Once you've gone through your entire facility identifying potential areas and checking out their capabilities, you'll want to make a utilization list. With the Emergency Management Committee or similar group review the list and try to match up the areas with what you would use them for. You will then want to set up guidelines as to when you would activate these areas and under what conditions you would activate.

Having completed this task, and again, dependent upon the identified use of an area, will it be possible to pre-stage supplies there? If you make that commitment understand that you'll have to track and monitor those supplies just like you would do with your main inventory. If this is not feasible, can you at least pre-stage supplies for your identified areas in one central location? By doing that you would have the benefit of a pre-established supply cache and the benefit of being able to monitor everything from one location.

Another option is looking at other areas off site that your facility may own or have access too. You would repeat the above process for determining their capabilities as well.

Remember to consider logistics when identifying off site facilities. How far away is the alternate facility? What are the roads and terrain like? Will you be able to transport equipment, supplies, staff and patients there safely during a disaster? Would it be safer to stay at your own facility?

These are just some of the questions you need to ask and answer before making a determination. What about using off-site locations in your community as an alternate care site? One suggestion before you start shopping around for facilities. Check with your local, Town, County, City Community Emergency Manager and office to see what areas they've identified, designated and what their intended uses are.

 This will save you a lot of time and effort on your part. If you do find a facility that's not pre-designated and would fit some of the needs of your facility, your work has just begun. You'll have to go through that same checklist procedure and you'll have to have an agreement in writing. Get your Legal Department involved as soon as possible on this issue.

Things like usage and liability really come into play here and must really be passed by your legal department. If the intended use during a disaster is agreeable to the facility owner, be sure to relay all of that information to the facility attorney. They in turn, will then draw up a contract or a Memo of Agreement (MOA) to be signed by all of the appropriate parties.

Two other options you could consider if the financial resources are available would be; trucked in mobile treatment units and portable medical facilities. Both of these options have upsides and downsides to them. Either of these options gives you the ability to expand services and/or provide them during a disaster.

They are pretty much self-contained and can be utilized as stand-alone units. They've been used across the country very successfully during disasters such as Hurricane Katrina, Joplin MO and Hurricane Sandy. But as with anything else there down sides to consider as well.

Aside from the financial cost and commitment, which can be considerable, you must think about whether or not the trucked-in unit would be able to get to your facility or designated area in a disaster. Regarding portable units, you'll need a storage area large enough for it and an appropriate area to set it up. It also takes a fair amount time and staff assistance to erect this type of facility.

My intent here was to illustrate some of your options regarding alternate site identification and use. I hope I've done that for you. Each facility and Emergency Manager will ultimately have to identify their own needs and resources.

We are now going to look at and discuss some of our basic necessities such as linens, food and water. During a disaster, certainly one of any consequence, intensity or duration, these items could literally be worth their weight in gold. Let's start with linens. Sheets, pillow cases, pillows, blankets and of course towels. Does your facility clean its own linen or does a vendor provide that service? Either way, how much do you or can you stockpile? What are your physical or budgetary restrictions? What about storage capacity? How many days reserve do you have on hand? How far away is the vendor and how long would it take to be resupplied? Exactly where the vendor is physically located can make a difference.

On Long Island, New York, during Hurricane Sandy in 2012 several healthcare facilities utilized a vendor located on the south shore.

During the storm the vendor was impacted and unable to provide service. This had the facilities scrambling. Where is your vendor located? Will that be an issue for you? If it is, what is your back-up plan? Work together with your Laundry Manager, if you have one or the person responsible for it in your facility.

Discuss ways of stretching supplies, like only changing sheets every two to three days if not visibly soiled. Removing pillow cases and using them as towels. These are simply suggestions and should only to be considered during extreme emergencies. But be creative and think outside the box with the rest of your staff. Obviously, patient care is paramount at all times. But these are the type of things that should start you thinking.

Does your facility have rehabilitation (Rehab), occupational therapy (OT) or psychiatric units? Are there any washing machines in those areas? Can they be utilized in any way to assist the facility during an emergency?

 Do some brainstorming, check with other facilities, and see what they are currently doing or what they have done in the past. As you can see, even something that would seem as simple as linens can open up a multitude of questions that need to be answered. I hope that you are able to find the answers for your facility.

Now let's talk food and water. We all need to eat. And at healthcare facilities we have two different and distinct populations; our patients (residents) and our staff. And along with that goes all of the dietary restrictions that people have and that we need to accommodate. Again, you'll need an assessment of your facility's supplies and resources.

Speak to the Food & Nutrition (Dietary) Director about what they have on hand. Find out if they have made any "disaster" plans for providing meals during a disaster period. Again, I was extremely fortunate to have a real professional in this position. Our Director, Theresa Giangarra, worked closely with me every step of the way. She provided inventory lists, meal plans and their implementation, vendor information and even staffing levels and needs.

She also provided a Department Emergency Management plan that was so well put together that I used it with her permission as a template for the rest of the facility departments. Without a doubt, she was one of the smartest, hardest working, most professional people I've ever had the opportunity and pleasure to work with. She also became one of my closest friends at the Medical Center where we worked. One of the things that was discussed was possibly going to a two (2) meal day for staff in order to stretch food stocks in the event of a long response. (This is one of the areas where the Joint Commission's "96" rule comes into play). At an Emergency Management conference hosted by the local Public Utility, I spoke with a vendor who had a commercial version of MRE's (military Meals Ready to Eat).

I brought several samples back to her to try. This is something that might be worth looking into for your facility. Our Director also provided food and drinks to the EOC during any actual activation for staff for the duration of the event. As it turned out we had the capability of providing meals for three days on full menu and up to five to seven days on modified menus. We also had a stockpile of five hundred (500) individual gallons of water to supplement patients and staff if necessary.

And when the expiration date of the water was nearing, we would rotate stock and allow staff to purchase the gallons for their own use. So it was never wasted. Since leaving the Medical Center where I worked, they have made several changes and additions to the food and water supplies. They now also stock additional baby food in their emergency management cache and also stock tube feeding supplies.

They've also converted from gallon jugs of waters to liter bottles (4000), so that it's easier to distribute and use for patients and staff alike. This is a good example of how you need to stay current and adapt to your changing needs and demands.

Let's continue speaking about water but shift gears from Dietary needs to facility needs. When you're doing your physical plant assessment with your Facilities or Engineering Director, make sure the both of you discuss your water supply. You'll need it for drinking, washing, sanitation, cooling and even firefighting. How is the water fed to the site? Is there more than one feed? Do you have a vendor that can bring water in (Like a water tanker truck)? If you are part of a network or healthcare system, will they be able to provide any support (water, food) to you from the unaffected facilities? Does the system you are part of have vendor agreements in place (MOU or MOA) that you can tap into for assistance? Again, much of what we do requires time and legwork. That's why networking and contacts are so important to you. They can save you a lot time and effort.

I'd like to segue into vendors and service agreements or "Memos of Agreement" (MOA's) and "Memos of Understanding" (MOU's) at this time.

First you'll need to determine what services are needed for the facility during a disaster that are not already provided in some way shape or form. After listing and prioritizing these, check to see (if you belong to a system) if a contract exists somewhere that you can tie into. If not, then you'll need to put something together yourself for your facility. Either way, you'll be dealing extensively with your facility legal department and attorney.

Explain to them what you need and they will help guide you through the entire process. Once you have these in place, make sure they are current and updated annually.

Also be sure to keep copies of all of them in your facility compliance binder, so that if questioned by the accreditation surveyor, you can produce them immediately.

Alright, for a moment, let's assume you've collected or purchased all of the ancillary emergency management supplies you feel your facility needs. Now what do you do with it? Well, a good place to start is by setting up a spreadsheet listing all of the inventory by item, quantity and location. This will give you a quick snapshot as to what you have and where you have it stored. One thing you definitely want to keep track of is, if any of your items have an expiration or use by date. If they do, make sure these items are up front and easy to get to in your storage area. Having an emergency management equipment and supply inventory is one thing, maintaining it is another. This is not something you can or want to put together once and forget about. You need to know exactly what you have at all times, how much you have of each item and is it usable. Make sure to check not only quantities, but to make sure *all parts* of an item are there as well. If at all possible, try and store all items in one central location.

During usage, and especially during a disaster response, you want to be able to get to the necessary items and distribute them as quickly as possible. If you, or someone else has to run around to three or four different locations, you're wasting precious time. And as you know, in a real time event it's the one commodity you don't always have. Make sure you preplan and use your time wisely. I'd like to make a suggestion here that could save you some possible headaches down the road. Have your equipment and supply inventory checked physically by someone at *least on a quarterly basis, but preferably monthly if possible*. Make sure whoever is attesting to the accuracy of the inventory is *signing off for it somewhere*. You want to have accountability when it comes to this. Keep a copy of that attestation in your Emergency Management binder to reference if necessary.

By performing these checks on a quarterly basis, you're less likely to be surprised in an unpleasant way. For example, if you had 50 portable cots listed and you went to use them, you don't want to find out there are only 37. Or that half of your twenty respirators are missing filter cartridges and are now useless.

I'm sure you fully understand the importance of having not only a well-stocked inventory of necessary items, but an accurate one as well. Keep in mind that these items are to be used for your first line of defense. You need to maintain them well.

I'd like to touch briefly on something that I believe, is misinterpreted by many of my healthcare Emergency Manager colleagues. And that is the "TJC" standard that address' what's commonly known as the "96 hour rule". The TJC EP does not actually require hospitals to stand on their own for 96 hours.

Rather, it expects facilities to determine on their own *whether or not they can survive* for that period, and if they can't, what steps will they take in a disaster. For example, if a facility determines that it could only sustain itself for eighty (80) hours, the logical follow-up at that point would be to evacuate. Many people I come across are working under the impression that they *MUST* be able survive for the full ninety-six (96) hours.

That's just not the case. What you actually need is a plan for functionality over that time frame. What you also need to do though, is set up a workable matrix that addresses all of your essential items, how much you have on hand, how long they are expected to last, and what do you do when you've exhausted your supply of an item.

Let's move on now to *training, exercises and drills.* We'll begin by discussing the importance of training, and why its effective use with staff can make all the difference during a disaster. This is true of all phases from preparation, mitigation, response and even recovery. An important thing to remember about training is that you will be dealing many different levels and abilities of your staff.

You need to be able to reach everyone in your facility. Even if they are functionally illiterate or even if English is not their first language. This takes some doing and is something that must be done diplomatically and professionally. I feel it's important enough to mention again, how important our staff is to our facility. We need to protect and educate them so they can function in their assigned duties during a disaster, without becoming victims themselves.

Let me ask you the following questions; do you personally provide any training for your facility staff currently?

Or, do you provide training modules and or informational material to another department to provide facility training? At our facility for example, I only provided personal training to the Public Safety Department. I was also one of the presenters for Human Resources (HR) for new employee orientation classes. Then I provided training modules to our Nursing In-service department and Environmental Health & Safety to use in their annual Environment of Care training classes.

This was done for all hospital staff. In this way, I was able to reach all of our facility employees at least once on an annual basis.

The training modules effectiveness was quantified by a ten (10) question quiz at the end of the module. Staff needed to score seven out of ten to pass. If they failed, they had to be re-tested. Everyone had to take these classes, no one was exempt (not even the CEO). But of course, accommodations were made to fit the schedules of administration. This training was also reinforced at our facility by something we instituted that proved to be very effective.

 I'm speaking of weekly "Environment of Care rounds". Our "Environment of Care rounds" functioned as a team (VP of Facilities, Environmental Services, Engineering, Emergency Management, Security, Health & Safety and Bio-Medical) would go to a different floor and area each week.

We used a check-off list to inspect the area, then at the end of our rounds we would call staff together to review our findings. We also gave staff the opportunity to ask questions or verbalize their concerns. We would then finish by asking staff several pertinent questions regarding things they needed to know.

This was more of a review and reinforcement of information already provided to them as opposed to testing their knowledge. This was something that worked out extremely well for us. This program was put together primarily by our Director of Environmental Health & Safety, George Araujo, without whose expertise, knowledge and assistance I would never have been able to accomplish the things I did. George, who worked hand in hand with me, was probably one of the most qualified but underappreciated individuals at our facility.

Training programs will be different to reflect the needs, priorities and identified hazards for a particular facility. Like healthcare facilities, they come in all shapes, sizes and complexity. Just don't overthink the process. Identify your training needs, identify your target audience and then commence with the education program.

Make sure that it is written at a level that everyone is comfortable with and that the information is straightforward. When formulating an exam, make sure that the answers will all have a clear cut response to the question. You don't want them to be ambiguous in any way for your staff. Your goal is to educate them and prepare them for their roles during a disaster. If you can do this, you'll be a successful Emergency Manager.

I'd like to take a little time here, to show you just how important and relevant good training can be. I'm going to tell you about a gentleman by the name of Rick Rescorla. He was a retired U.S. military Officer of British birth who had served with distinction during the Viet Nam War. He originally went to work for "Dean Witter Reynolds" in the world Trade Center in 1985.

He then went on to become Vice-President of Security for Morgan Stanley 1997 when they merged, also at the World Trade Center. Rescorla was often at odds with his bosses about what he felt was important training (evacuation in particular). He received a lot of push back over the amount of drills and degree of participation he expected.

In 1990, he asked a good friend, Daniel Hill, whom he had served with and who was trained in counter-terrorism to come to New York and visit the World Trade Center. Rescorla requested that Hill perform a Security Assessment of the building. He also asked him what would be the most likely route of attack if he were a terrorist. After being able to enter the building and make his way all the way to the basement parking garage without being challenged or stopped, Hill felt this was the easiest access. This, along with the fact that the main structural columns were exposed, made them a tempting target for a vehicle laden with explosives.

That same year, Rick Rescorla and Daniel Hill sent a report to the Port Authority of New York and New Jersey, the official owner of the building. In their report, they pointed out the need for increased security and surveillance for the parking garage in particular. Unfortunately, these recommendations, were never acted upon, probably due to their high cost. This was noted in James B. Stewart's biography of Rescorla, *Heart of a Soldier.*

Of course we all know what happened in 1993 with the World Trade Center. Needless to say, after this, Administration was somewhat more compliant with the training requirements since Rescorla gained credibility, stature and authority after the bombing.

Even with this, it was still a constant gripe of Administration about all of the "interruptions" that Rescorla required. But he kept at it and stayed the course.

On Tuesday, September 11, 2001, that training proved to be invaluable to the people of Morgan Stanley. There were 2,700 hundred employees working that fateful day. As a result of his commitment to training, his perseverance and professionalism and personal leadership, 2,694 employees made it to safety that day and returned home to their families. Unfortunately, Rick Resorla and five others did not. Not being able to account for those final five, he returned to the building for what would be his final time.

I wonder how those Administrators felt about training the following day on September 12th while they were safe at home. Never, never underestimate the importance of good training. It may just save your life someday.

At this point we need to take some time to discuss *"Drills and Exercises"*, their differences, uses and benefits. The difference between a drill and exercise is as follows; a drill focuses on one specific task, while an exercise covers all phases.

To show it in the form of an analogy, I'd like to use an example from a very good friend, colleague and mentor (Francis L. Melton, Texas A&M University, and TEEX). When a football team practices, they run *drills*; passing, tackling, kicking etc.

When the whole team scrimmages, they're running an *exercise* of all of those tasks together at the same time. This is a great visual to use when explaining the difference to others. Another great visual is Francis himself. He stands close to six foot eight inches tall and has a voice that's a cross between the actor James Earl Jones and the late singer Barry White.

When he presents, people tend to listen. But even with his imposing stature, he has the ability to connect with his audience and puts them at ease immediately to make the entire learning process enjoyable. And that's an important point to remember. This is especially true when teaching adult learners.

For the healthcare profession, this might actually translate to a conducting a specific unit evacuation drill or a fire drill. Or perhaps conducting a notification drill to see how long it takes senior staff to call in when notified. You could drill their response time for responding to your Emergency Operations Center (EOC). Hopefully this clarifies the difference in terminology for you and better illustrates the meaning.

Now we'll look at some the different types of exercises and how utilizing them, can and will benefit you. The three different types of exercises that I would like to focus on are; *tabletop exercise, functional exercise and full scale exercise.* While there are other types, I feel that these three are the most pertinent and relevant for the healthcare emergency manager, so these are the ones I'm going to focus on. Just on principle, many people dismiss using tabletop exercises for their facilities. The primary reason for this is because the accrediting organizations that survey healthcare facilities do not give credit for these types of exercises.

As a result many people don't take the time to utilize them. On the other hand, I personally think tabletop exercises are an excellent tool to make use of at your facility. I averaged approximately four tabletop exercises annually. All of them benefited the facility and our emergency management program in some way.

This, was in addition to the functional and full scale exercises we participated in, as well as any actual activations of our Emergency Management Plan. I specifically liked tabletop exercises because they were low key, low stress situations for the participants. They were also low cost and time efficient. I would pick a specific topic or issue and we would run through it in fifteen, twenty or thirty minutes. Sometimes they would be part of the Emergency Management Committee agenda. The benefit was that it got people thinking about emergency management issues on a fairly regular basis.

They then became somewhat proficient in its terminology, concept and practice. And as a bonus they become more confident in the entire process. As a result, when we had actual activations, staff were much less tense during their response. They were more comfortable in their response roles within their departments as well as the facility.

I also made sure to write up an "After Action Report" (AAR) and document all of our actions during the tabletop exercise. I then included it in my compliance binder along with all other exercises and actual responses. During the survey process, the surveyor more often than not asked why we conducted so many tabletop exercises since we weren't receiving any credit for them. After I explained it to them, they would usually wonder why more people didn't use them for that purpose. I don't really see any downside to utilizing this tool and would strongly suggest you give it some consideration. It shows the surveyor that you're going above and beyond the basic requirements of their standards.

*Functional Exercises* on the other hand do receive credit from accrediting organizations. They are more labor and time intensive and thus require more planning and participation.

A functional exercise will normally simulate an emergency in a realistic manner if possible. It may even include the influx and use of "moulge victims". As its name would suggest, its goal is to test or evaluate the capability of one or more functions in the context of an emergency event. By its very nature, these exercises are more intense and more realistic with their response of personnel. They tend to have a much higher stress level, especially if the exercises' implementation has not been pre-announced at the facility.

The functional exercise makes it possible to test the same capabilities and responses that would be tested in a *Full Scale Exercise,* but without incurring the associated costs or safety concerns. The functional exercise is well-suited to assess things such as; direction and control of emergency management, communication and information sharing among organizations and allocation of resources and personnel.

These type of exercises if done properly, will take time and effort to put together. You'll be assigning roles to players as well as monitors. You'll be dealing in "real time" and staff will have to respond to messages delivered to them during the exercise.

Their role is to work through these "problem messages" and adapt to the ongoing situation as if it were an actual emergency. Make sure you don't try doing it alone or in a vacuum. By this I mean making sure the Emergency Management Committee staff are aware of and participate in the process. You will also need to keep Administration informed of what your intentions are, what you'll be exercising as well as when you plan to exercise. Please keep in mind what I stated earlier about keeping Administration informed and maintaining them as allies of the entire process.

This applies as well to the entire Emergency Management Program. The better informed they are the more comfortable and responsive they will become.

Make sure to fully document everything about your *Functional Exercise.* In addition to your formal after action report (AAR), conduct a critique and a debriefing of staff involved in the process. This will give you a good overall view of everything that transpired during the exercise. Also be sure you list any shortcomings or deficiencies identified and how you intend to address them before the next exercise. Have a plan of action in place with a realistic completion date for fixing the identified issue or issues. If you are unable to remedy the identified issues prior to the next exercise, at least update and address it in the after action report for that exercise. Otherwise, the surveyor may pick up on your inaction and call you on it. This is not a position you want to find yourself in. It's also the prudent and appropriate thing to do.

*Full Scale Exercises* put you in the major league. The full scale exercise by definition, includes all the components of a functional exercise and adds the addition of actual responding field units.

It is intended to test and evaluate the operational capability of your entire emergency management program with other agencies in an interactive manner. If conducted properly, a full scale exercise is as close to the real thing as possible. One of the main functions of a full scale exercise is to test the coordination of all participants. These usually range from the healthcare facility, to local, county and state Emergency Management representatives, local government, utilities, Police, Fire and EMS personnel.

A *Controller, Monitor and Observers* will be assigned to participate in this type of exercise as well. This will allow for an overall objective appraisal of the exercise. A full scale exercise will also make full use of your "Emergency Operations Center" (EOC) and the "Hospital Incident Command System" (HICS). While full scale exercises offer the most realistic response outside of an actual event, there are issues to be dealt with. There is an extensive time commitment for design and implementation for this type of exercise.

And even though your healthcare facility will only be a participant, it will still be time intensive for you, especially depending upon the functions that will be exercised. Then there is the additional cost in real dollars for personnel participation and resource allocation costs.

And finally, the increased safety risks that you must consider as well as potential liability concerns. Staff will be acting as if to an actual disaster. You need to educate and protect them.

This is a major undertaking and should never be entered into lightly. However, if conducted properly it affords the best opportunity for responding to disaster without it being an actual event. In my mind, it is well worth the time and effort to have the opportunity to work alongside other community members and agencies in this type of setting.

Something to keep in mind when running any kind of drill or exercise, is to make sure that they don't ever become routine. Don't run them on the same shift all of the time or on a particular day of the week or in the same timeframe. If staff know when every drill or exercise is coming, they won't respond in the way you need them to. There is always a different mindset between practiced events and real events.

Shake things up a little if you can, run them on off shifts or on a weekend if possible. See what kind of response you get. This is not in order to make anyone look bad, but it might give you a better picture on what shortcomings you have and what you need to do to plug any gaps identified in your response plan. Better to find them now, rather than during an actual disaster response. On the opposite side of the coin, you don't want them too realistic or a total surprise either. That can actually backfire on you.

Let me illustrate with an example; in a total surprise exercise wherein a staff member was "taken hostage" the nurse who said she was not warned that she would be threatened and taken hostage by a gunman during an emergency preparedness drill is suing her former Colorado nursing home employer for compensatory, punitive and actual economic damages, claiming distress and mental anguish. Although it didn't happen in this case, I can see several extremely bad outcomes with this particular drill and how it was executed.

What if there was an armed off-duty Law Enforcement Officer present or someone else with a weapon and drew their weapon in response? And what about if someone had a heart condition and had a heart attack as a result. It's also not a good idea to run surprise unannounced drills and exercises without your boss knowing about it either. Remember what we said about not trying to make anyone look bad.

 Make sure you use some common sense when putting these drills and exercises together. It will save you some heartache later on.

The next thing that I would like to discuss that ties directly into drills and exercises is the *"Hospital Incident Command System"* (HICS). Let's begin with a little bit of the history behind "HICS".

Incident command as we know it today had its genesis from the California wildfires of the 1970's. Multiple agencies would respond but would be ineffective in coordinating their efforts as a result of proprietary radio codes and terminology.

Studies determined that response problems were often related to communication and management deficiencies rather than lack of resources or failure of tactics. Seeing the need, civilian ICS was built based upon the management hierarchy of the US Navy and it was originally used mainly for the fighting of wildfires in California and Arizona.

From there, it made its way over time into all aspects of public services; police, fire, EMS, and eventually healthcare facilities. Originally, healthcare facilities utilized the "Hospital Emergency Incident Command System" (HEICS) for their response to disasters situations.

Since its inception in the late 1980s, the Hospital Emergency Incident Command System (HEICS) served as an important foundation for the more than 6,000 hospitals Nationwide in order to prepare for and respond to various types of disasters they might encounter.

When the fourth edition of HEICS was developed, the value and importance of using an incident management system to assist as well with daily operations, preplanned events, and non-emergent situations became apparent. As a result of this, HEICS IV has evolved to become HICS; the *Hospital Incident Command System*.

As a scalable, comprehensive incident management system HICS could now be used for both disaster and non-disaster situations.

This has given much greater flexibility to healthcare facilities and with it, created a management tool that would be much more effective in its application. Remember, you can plug in what you need depending upon the situation.

You don't have to set up every position, unless that is what you need or want to do. An additional position that we added to the Incident Commander Section was that of EOC Manager. While not in the HICS guidebook, we found its application extremely useful and beneficial. This individual would coordinate all activities within the EOC and report directly to the Incident Commander. This allowed everything going on in the EOC to flow much more smoothly. This is simply a suggestion, but it has worked very well for us. This position would also monitor the timeline, make appropriate notifications and make sure everything was being documented.

This could be something you might consider utilizing to make things easier for you in the EOC.

As Emergency Managers I'm sure all of you have a firm understanding and grasp on the concept of "Incident Command". I wouldn't presume to lecture you on the basics or even the application and benefits of its use. Suffice it to say, that with its structure, span of control and unified command concept, HICS has become an invaluable tool for us to utilize. If I can digress a little bit, HICS is an ideal topic to be used for a tabletop exercise. It gives you a great opportunity to familiarize staff with their assigned roles and the duties expected of them, but in a low stress setting.

With the newest version, the fifth edition just being released in 2014, there have been some important changes.

There were also additions and deletions to the titles, structure, guides and forms of the previous edition. To review and familiarize yourself with all of the changes go to the Hospital Incident Command System (HICS) Guidebook. From there, just click on the Guidebook for a downloadable copy. Even if your facility is not currently using it, this will give you the opportunity to get up to speed on the changes. You can find this online at; www.emsa.ca.gov/disa.

The next topic we'll touch on, is the proverbial "800 pound gorilla" in the room. A *"Full facility evacuation"* is something we all have to consider and even train and prepare for. It's not something to be considered lightly. Primarily, because there are so many variables to consider. These all have to be considered before you pull the trigger on implementing a full facility evacuation.

If it something like a hurricane approaching, you'll obviously have some planning and preparation time. But even with that, where will all of your patients go? Can you even physically move all of them without putting their health at even greater risk? That has always been a major concern for long term care facilities. These facilities know that this type move while saving lives, more than likely cause a few to be lost as well.

Will your alternate care sites be capable of handling this kind of patient census and care needs? What about other area healthcare facilities? Can they take your patients, or will you all be competing for the same space somewhere?

The thing of it is, is that this can be done, and has been done successfully on many occasions, by many different types of healthcare facilities. I just don't have a simple answer for you or a standard plan that I could share with you. If I did, I would certainly share it with everyone.

In healthcare, we're in the business of making people feel better and saving lives. Sometimes, evacuating an entire facility is what is needed to accomplish this specific goal.

This though, is one of those things that truly must be tailored to a specific facility's needs. What kind of resources will you need to accomplish it? Staffing, equipment, vehicles to name but just a few. This plan, more than anything, needs the input and knowledge of key facility staff participating in the process from inception all the way through implementation. You'll also need to include the appropriate outside agencies in the planning process as well, if for no other reason than to ascertain what community resources are available to you and your facility. Hopefully, it will never be something that you have to implement for your facility, but you do have to prepare and plan for the possibility of it.

One of the main concerns when performing a full facility evacuation in either a "Long Term Care" facility or a "hospital" is the potential for patient injuries and/or fatalities. There are even some industry estimates for a 10% fatality rate for LTC residents.

In a high rise facility, such as one that is five stories or higher, there will be additional concerns. Across the country the overall workforce is aging. People are working longer thus at an older age. This will have a direct bearing for you when putting your full facility evacuation plan together. The UN predicts the age group over the age of 65 will double within just 25 years. Unless we see some sort of extraordinary advances in medical science over the coming decades, we can expect that this ageing population will continue to experience the same diminished visual acuity, depth perception, reduced hearing, loss of the sense of smell, as well as a higher prevalence of people with mobility impairment as they age. All of these things impair an individual's ability to evacuate quickly during a disaster or emergency. In the future, "fire codes" may even have to be revisited and updated as far as calculating evacuation times when a fire escape stairs is full, taking all of these things into consideration.

In fact, we only need to look back to the evacuation of the New York World Trade Center Towers in 2001 to see examples of how some of the occupants found the egress route via the stairs difficult. It has even been estimated that approximately 1,000 of the 9,000 surviving occupants had some form of impairment or activity limitation which restricted their ability to via the stairs.

Again, this all becomes much more critical in a high-rise facility. My facility's main building was a twenty story tower. This posed some unique issues for me when planning my own full facility evacuation plan. I worked closely not only with Nursing and Engineering, but with the local Fire Department as well. I'm fortunate in the fact that we never had to activate the full plan.

I would like to take some time at this point to discuss the importance of comprehensive and accurate documentation. I can't stress it enough and I'm sure any Emergency Manager that's worth their weight understands this concept only too well. Documentation is woven through every aspect of the Emergency Management process. It starts with the documentation that we put into our response procedures, our Emergency Operations Plan (EOP) all the way through the entire Comprehensive Emergency Management Program (CEMP). It can also be found in our Continuity of Operations Plan (COOP).

"Continuity of Operations" is an effort by the facility to ensure that Primary Mission Essential Functions (PMEFs) continue to be performed during a wide range of emergencies, including localized acts of nature, accidents and technological or attack-related emergencies. In essence, during any and all emergencies or disasters. From the planning phase, prevention, mitigation to the response phase, recovery and finally evaluation and improvement. It's the thread that brings cohesiveness to it all.

It aids us in drills/exercises, resource allocation, inventory, networking, vendor commitments, patient admission, payroll and restocking to name just a few. But maybe most important of all for the facility's bottom line is that without it there can be no reimbursement of State or Federal funding. And again, depending upon the disaster, reimbursement could run into the millions of dollars, not only for repairs but for mitigation efforts as well.

You would not want to lose any of that as a result of inadequate documentation. I'm not going to drag this out, especially for many of you, as I'm already stating the obvious. But rest assured, good documentation can make all the difference in an Emergency Management program as well as during an actual disaster response.

The final topic in this chapter that I'd like to speak about in relation to preparedness is the *"Hazard Vulnerability Analysis"* (HVA). By and large, this is one of the most important and useful tools we have to work with in healthcare emergency management. First of all, I'd like to clarify an important point for you before you prepare your HVA. Often times in Emergency Management and other related disciplines there is confusion about the difference between the terms and meaning of *"Threat and Risk"*.

On occasion, even experienced professionals will sometimes having difficulty explaining it. But it is important, especially when preparing your "HVA" to know the difference. Let's look at some definitions.

*"THREAT* "can be defined as; *"An individual or group with the capability and intent to cause damage/harm to people and/or property".*

The "HVA" is simply a threat assessment tool for us to use. We conduct our threat assessments using the "HVA" to evaluate human action that can adversely impact on individuals, critical infrastructure, business continuity, operations or assets.

*"RISK"* can be defined as follows; *"The vulnerability of your staff and facility to a recognized "THREAT" activity and the likelihood and consequence of it occurring".*

Simply stated, "Risk" is what we face should a "Threat" be carried out. It's why we prepare for disaster and other emergency events. It's also why we prepare the Hazard Vulnerability Analysis (HVA).

The HVA is the tool that we use from the very beginning of our planning strategy. If you remember in chapter one where we try to identify the potential hazards or disasters that our facility may face, this is the very tool we would use to perform that task. And like almost everything else in healthcare emergency management, this is something that as the Facility Emergency Manager you don't want to try to prepare alone. Utilize your emergency management committee or environment of care committee and the expertise and knowledge of your facility staff to engage, participate and complete this very important function. This will actually make for a better HVA and hopefully a strongly response.

The HVA, if done properly, is something that should not be rushed through. Think about the hazards you'll be identifying. Then think about the potential risk that each one poses and how prepared you are both internally and externally for a facility response.

As a group and as the Emergency Manager, always be frank and honest when grading yourself. Don't ever gloss over your true preparedness level as you'll be doing a disservice to the facility and yourself. If you overate your facility you'll only pay for it on the backend with a poor disaster response. That could result in loss of property or injuries. And that in turn, it might even lead to you seeking new employment. None of those are really good options at all. So do yourself a favor, keep it honest and learn from it.

Let's talk about what the HVA is and what it provides for you. The HVA is one of the very first steps we take regarding preparedness and in putting together our *"Emergency Operations Plan"*. The HVA serves the facility primarily, as a needs assessment tool if you will, wherein, you can identify hazards, determine probability of occurrence, severity of impact, potential risk and preparedness level.

By reviewing your grading, it will give you a quick snapshot as to your readiness level. Something to also keep in mind is that "The Joint Commission" and "NFPA 99" both require a hazard vulnerability analysis (or assessment) to be completed and updated by the healthcare facility. There are plenty of HVA models and templates that can be utilized.

The one that I've come across the most in my travels is the "Kaiser-Permanente" version. I used a modified version of this template myself at our facility and it worked extremely well. It is well thought out, presented and easy to use.

While I don't make specific recommendation or endorsements, the fact that this version is so widely used around the country speaks for itself. At the very least, take a look at it and see if fits your facility's needs. Or adapt it to fit your facility's needs.

What this template does, is take the Natural, Technological and Human Hazards along with Hazardous Materials and breaks them down on separate sheets. It then applies the grades you assign for probability and weighs it against the severity of a given hazard. It does so, with the following; human, property and business impact along with preparedness, internal and external response. It then automatically calculates your aggregate score for the final column which is risk or the relative threat of a hazard. By looking at the final scores, you can easily identify the highest potential risk associated with your facility. These will then be the ones you key in on and exercise on.
They will also be the ones that the accreditation surveyor will be focusing on as well. Finally, with this version there is a summary page which will also allow you to graph your results. Thus, making it easy to reference.

Whichever HVA template or tool you decide to use, remember that after the initial completion, it must be updated on a regular basis. That schedule can be up to you, but I would strongly suggest that you update more than once annually.

That is the bare minimum and doesn't really provide a good continuity in your preparedness process. You should update after any actual response, especially a large one, after any major exercise on an identified hazard and at least every six (6) months. This will help you and the facility remain current, updated and prepared to face both identified and unknown hazards that are out there. And, as with the initial preparation, updates should also be conducted in a group session.

The greater the input into the updating process the greater the final result will be. In our profession, as with many others, good and improved results are what we are always striving for. I hope that the topics we've covered in this chapter as well as the suggestions I've made here will benefit you as a Healthcare Emergency Manager.

Ultimately, you must sort through the various tools and determine which ones will work best for you in your mission and at your facility. Take some time when performing this task as there are many good resources out there. Again, reach out and see what your colleagues are doing in other facilities and don't forget to check the various blog sites or professional associations for assistance. This will especially help newcomers to the field and will keep veteran ones current.

Another important concept to keep in mind is that the whole community prepares, responds and recovers together when facing a disaster. You see this all over the country but especially in small towns and cities where the sense of community is very strong. Nothing happens in a vacuum. We need to be there for each other. Healthcare facilities are a major part of that community infrastructure. In our next chapter, we will be moving into the response phase and looking at how the healthcare facility which is an important community asset becomes an even greater one during times of disaster.

# CHAPTER 4

## Responding to a Community Disaster

You're tracking the weather channel as a category 3 hurricane approaches with predicted landfall in your area. Or after severe thunderstorms and hail, the tornado warning siren just sounded for your town. Flood waters have risen and are inundating the roads in your community. Wildfire smoke and flames are threatening to engulf your area. A train derailment and toxic release has just occurred within your city limits. These are all potential, realistic events that can and do occur. What do you need to do as a Healthcare Emergency Manager to respond? With your community bracing for the inevitable, you and your facility need to ramp up preparations to accommodate both the patients that are within your facility and the potential medical surge you may soon be receiving. The fact is, your facility is an integral part of the community infrastructure and as such that community will dependent upon your response and support during this time of crisis. Keeping that in mind, the big question that needs to be answered is, where do you start?

For the purpose of this chapter, and because of past historical data, I'm going to use the hurricane scenario to highlight what I think are some of the essentials you will need to implement for an appropriate response. With today's meteorological advances in technology, observation and tracking we can pretty much start our planning at the four day (96 hours) mark. At this point, you should be meeting or planning to meet shortly with all of the appropriate department heads in your facility.

Ideally, by going through administration, who'll you'll need to keep fully informed in any event, you should request a facility-wide meeting. This way it gives you an opportunity to brief all of the key facility staff at once. Initially, you can also go over the facility's preparation plans and expected response timetable. Keep in mind, that the more information you provide to your staff, the better they will be able to respond to the disaster. You're the coach of the team as it were, in this case. So keep everyone informed and on the same page. Good communications start with you.

Assuming you are having a facility meeting, and after you've laid out the facility's plan of action, find out what the level of preparedness is for each of the department's attending. What are their supplies and inventories like? What if anything do they need? What are their current staffing levels? Not only full time equivalent positions, but are staff out on vacation, worker's compensation, maternity leave, or family medical leave act. If it's the summer (hurricane season) a department may have up to a quarter (25%) of its staff out on vacation. This can and will have a direct impact on the facility's ability to respond. This is especially true if the event's duration and impact will be over a period of at least several days.

At this point, consider opening and activating your EOC if possible. Even if it is just one or two people answering phones and monitoring progress from your local Office of Emergency Management. This way, as things get more intense, the EOC is already functioning. All you'll be doing at that point is adding staff and ICS positions. This also is the point in time that you want to start reaching out to your supply and equipment vendors, reviewing any contracts or MOA's as well as other area health care facilities. If you're a system hospital find out what support you can expect if you are in the affected zone.

At the three day (72 hour) mark, your facility should be anticipating patient discharges to free up as many beds as possible. Remember, not only are you likely to receive victims from the storm, but you'll have to deal with the worried well presenting themselves at your facility. Another challenge you may face is from the "special needs" population within your community. These are people who are home but may have a chronic medical condition, or are semi-ambulatory or have diminished mental capacity. These individual's either on their own or with their families may show up on your doorstep. You need to be aware of this and plan accordingly well in advance of any event. If you don't, your emergency department and or your facility could be overwhelmed before the storm even hits. If that happens, the community loses an integral part of their community infrastructure that they are depending on. Don't let that happen. Work together with your Emergency Department Director as well as your Admissions Department Director. Monitor both ED admissions and bed availability throughout the facility. Also make sure you, or the staff in your EOC are receiving all current updates from your local EOC. In turn, keep your own staff informed as much as possible regarding your facility's response actions.

Your facility should also be considering the suspension of elective non-emergent surgeries. The final decision does not have to be made now, but it should be on the front burner. You can probably hold off until about thirty-six (36) hours prior to the hurricane making landfall. Even then, the cases need to be reviewed on an individual basis for need and ability to discharge or keep the patient through the storm. Department heads should be informing you as to their readiness. Have they set up appropriate staffing levels to get through the storm period?

Between the three day (72 hour) and two day (48 hour) mark consider when you would activate any "alternate care sites".

Whether on site or off site, can you staff and supply them if necessary? What would be their use to you during this event? Do you have areas set up for staff to sleep, shower and eat during their off hours? Does your facility have the authority to mandate staff to remain at your facility during the storm? What about staff hour changes from 8 hour shifts to 12 hours or 16 hours? These are some of the things your facility should be working on and trying to resolve at this time.

At the 36 hour to 24 hour time frame your EOC should have several additional ICS positions staffed. Each facility needs to decide which ones they'll be. Information should be flowing steady between your facility and your local EOC constantly now. At this time, they should have a good idea as to what your needs and capabilities are, how many beds you have available, and what type. Also, you should be letting the local EOC know if there are any specific needs your facility has and request their assistance in resolving them. There should be no surprises for either side at this time.

Twenty-four hours ahead of the hurricane landfall, all building exteriors should be checked, any outdoor furniture or planting pots, refuse containers or loose items should be removed or secured. All building windows should be closed and secured as well. All facility vehicles should be gassed up and secured if possible. Most Municipalities will pull emergency vehicles off the road once the winds reach a sustained strength of 55 miles per hour. You don't want to risk your staff or equipment either. If you have two-way radios these should be assigned at this time to Administration, Engineering, Maintenance, Housekeeping, Nursing, Emergency Department and anyone else you feel appropriate or have radios for. All departments at this time should be in disaster mode.
At the twelve hour mark, your EOC should be fully functional with all appropriate staff assigned. Your facility should be in

continuous communications with the local EOC. At this point, it's now just a matter of waiting for the storm to hit. So, the hurricane makes landfall as a weak category two storm. That still means winds of around 100 mph and heavy rain and flooding in low lying areas. The storm was fast moving so it is already through your area. But in the time it took to go through your community, it has left its mark both on that community and on your healthcare facility. What do you do next?

You'll need to begin with a full facility assessment as quickly as possible. You should have been receiving reports from various departments to your EOC throughout the storm. So, hopefully, this should not take a lot of time to assess your current status. Did the exterior of the facility suffer any damage? Is there any flooding within the facility? Are there broken windows, are trees down on your campus? Will that damage impede operations in anyway? If so, how soon can it be repaired and at what cost? Did you lose power, are you still on emergency power, and for how long? What are your current staffing levels and do they need to be adjusted? These are just some the questions you'll need answers too before you can fully formulate your community response actions.

Once you've sorted through this process, you can then begin to focus on the community at large and their needs. Have your liaison officer (HICS, ICS) contact the Local EOC and see if they are planning on transporting any victims to your facility. If so, try and find out the number of victims and the types of injuries that they will be presenting with to your Emergency Department. This will give you a little lead time to set up for the potential "medical surge" you may be confronted with.

If you can, also try to find out if any of the other healthcare facilities have suffered damage and as a result will be sending patients to you. If yes, try to make sure that the patient charts, medications and depending on the number of patients, medical staff as well. This will assist greatly in the efficiency and smoothness of the transfer. It will also provide a much greater continuity of care for the patient.

During Hurricane Sandy, our facility received approximately forty patients from a hospital that got flooded out on the South Shore of Long Island. The staff that came with those patients continued to provide care for them at our hospital. In the initial stages this proved invaluable as it prevented our staff from being overwhelmed and gave some additional comfort to the patients as they were dealing with familiar faces. An unforeseen issue arose within a day or two of the transfer. While all the medical staff that accompanied the patients were qualified, professional and licensed, they were not in our system. This precluded them from ordering medications, making electronic patient chart entries, picking up doctor's orders etc.

The Information Technology Department then had to adapt to come up with a system of temporary access codes that could be assigned and distributed to these staff members. And they did a great job. Human Resources also needed to develop a temporary ID card for them and Payroll had to develop a temporary status as well since the other hospital had closed. While it was confusing at first, everything did work out. Having illustrated this issue, try to plan ahead so that you are not caught up in a similar situation. Preplanning will definitely make your life easier. We were also one of two Long Island hospitals that a National Disaster Medical Service (NDMS) Disaster Medical Assistance Team (DMAT) was assigned too. This posed no issue at all, as they worked alongside of our staff and were able to act independently as well. They were fully functional and self-sustaining. Once activated they are paid as federal employees.

Having completed your own self check, your focus will now turn to your larger role of providing care and assistance to your community at large. Everything you do from this point on while benefiting your facility will also ultimately benefit the community. Obviously, the main function of your facility will be to receive and treat victims from the storm. While this is an important one and one that you excel at, there are so many more things that you and your facility can do to support your community.

For example, since hurricane season covers all of the summer months, how hot is it outside after the storm. People have probably lost electric power and could be suffering. Do you have the capability and space to set up a "cooling station" where people can come for a while and sit in an air conditioned environment. This is especially important for the elderly and small children (who would have to be accompanied by an adult). Conversely, if an event happened in the winter month's people would need someplace warm to go as well. These are just some of the ways you can continue to support your community in time of need. And since people in the community may be without power, what about their medical devices? Many have battery backup but will need to be re-charged. That is another service you can provide. Even allowing people to charge their cell phones so that they can maintain communications is an invaluable way of assisting your community. Can you provide simple nourishments to people from the community who may just need a few hours of shelter at your facility? Something as simple as a bottle or cup of fresh water can many times make the difference.
This is where those contacts come into play once again for you. If you are part of a healthcare system can you get some

supplies from the unaffected facilities or from a central warehouse? Can you contact the Salvation Army and see if they are willing to set up an aid station at your facility for the community? By using the facility as the focal point they may actually reach more people. And what about your resource list with those big warehouse stores that we mentioned earlier? Can they provide, in this time of need, any food or drink products that could be distributed at your facility to the community. Many times these companies are set up to do just that during times of community crisis.

And while the "good press" from such a gesture actually helps their bottom line, many do it simply from an altruistic view on wanting to help their community. Remember, most of the people that work at these large stores live right here in the same community that was impacted. They're our friends, neighbors and family. As such, it should be one of main goals to assist them wherever possible.

Information is the lifeline for everyone during a disaster. It's certainly no different for healthcare facilities. If anything, it may even have greater importance. So one of the things we need to maintain both internally and externally during this time is good and accurate communications. Internally, we want to be keeping our staff as informed as possible on the capability and status of the facility. By keeping staff well informed, they'll have peace of mind, thus allowing them to continue working in an efficient way. Also make sure, once information is received from the local EOC that you keep staff informed on the conditions of the community as well. They can't function in a vacuum.

They live and work in that same community. Things like, what areas may be flooded, which roads are impassable will keep them in touch with what's going on in the outside world. You should also consider what information if any, you can release to your patient population. They'll be worried as well for the same reasons as staff. I'm talking about basic information here, not detailed reports. Things like general flooding information or whether or not the community has power. You don't have to go into more detailed data than that. Too much information to the patient population could actually end up having an adverse effect.

External information on the other hand, is a completely different story. You and your facility will need as much detailed, accurate and up to date information as possible to fully function efficiently as a community asset. At this stage of the event make sure you have someone fresh and knowledgeable in their EOC role as liaison officer. This position becomes even more important now than it was earlier when the storm or disaster was approaching. That information received from the local EOC will help you determine what immediate and on-going tasks you'll need to accomplish.

Something else you could provide during the recovery phase of operations with the external information you receive is a "patient directory". Depending on how destructive the disaster was, there may be many victims. Not all of them will be brought to your facility (normally). Family members and friends may come to your facility looking for loved ones. By having a list of your own patients and well as victims brought to other area facilities you can provide information and hopefully peace of mind to some of your community residents. Again, networking helps immensely here.
You would simply be letting them know a certain individual was at a particular facility. No other information would be given so as not to violate any HIPPA regulations.
Remember to keep in mind that you'll be working on two fronts; maintaining the existing operations of your facility for

your current patient and staff population and providing expanded services to the community. How quickly you can get back to normal operational status will help determine how fast and to what extent you can provide those expanded community services we've already mentioned. This then, should be your next goal. Notify the local EOC of your current status. Ascertain what additional role they might need you to perform and confirm that you can provide it for them. At that point, you'll need to marshal your resources and staff to accomplish the task at hand.

Once you've done this, you'll want to review your staffing levels and schedules. You don't want your staff to "burnout". Any individual is only efficient for a given number of hours. After that, they start operating on pure adrenaline. That's when mistakes can happen. After all of the hard work that's been done by you and your facility staff, that's the last thing anyone wants. Make sure departments are monitoring their staff. They are our biggest resource and we need to protect them. They must be given rest periods. Make sure they do, even if you have to order them to take one. This is especially true for your Emergency Operations Center (EOC) staff. You need to have good "delegation of authority "and "orders of succession" plans set up prior to the disaster to accommodate a long term response and recovery. You will need to make sure that the positions are redundant by two to three deep with back-up individuals and don't forget to also address the off shifts as well. You don't know who will be available or be able to get to the facility at a given time.

As you continue to respond to the facility and community needs, it's important to remember that your recovery actions are also starting now. They'll run simultaneously along with all of your response actions. So you will definitely have your hands full, as will everyone else. Take a deep breath and move on to your next identified task.

If your facility has been unaffected by the storm/disaster or has received minimal damage you need to gear up for the influx of patients that you will be receiving. Hopefully prior to this as part of your Emergency Operations Plan (EOP) you have a designated area to use as triage for the surge. After all of this, you don't want your Emergency Department to become inundated and overwhelmed. As you receive, treat and or admit the victims it's important for your staff to assure these patients that everything will be alright.

The psychological trauma that many of these victims may have sustained should not be overlooked or forgotten about. Even during a crisis and disaster, a smile, kind word or a hand held can put people at ease and make the whole situation seem a little less foreboding and threatening. This also holds true for your staff as well. Depending upon the type of disaster encountered, there could be community devastation, loss of homes and or even fatalities. Friends and or relatives of staff may be among those lost. Secondary Trauma should never be overlooked or taken for granted. "Secondary Traumatic Stress" (STS) and "Post Traumatic Stress Disorder" (PTSD) among aid personnel and first responders is on the rise, just as it has been with our military personnel. Make sure that you have services in place and set up to deal with the emotional load that your staff may also be carrying. This should always be done ahead of time. Even though they are licensed healthcare providers, they are still people and human beings first.

They are, mothers, fathers, sisters, brothers etc. This is their home and community and when the crisis has passed they will need to deal with the aftermath of it all. This will certainly include what they've seen and treated at your facility. One of your main goals as a Healthcare Emergency Manager should be making sure that every staff member who arrived at work during the crisis, gets home safely.

As you receive and process victims from the disaster, you are also continuing to function in your primary role of healthcare provider to your existing patient population. Dependent upon the type of disaster, severity of impact and duration, your facility may have increased or decreased services on a temporary basis. Now is the time, once the immediate threat is past, to ascertain how much of a normal routine you can reinstitute. This will help not only the current operations of your facility, but will aid you and your community in the recovery process as well.

Have someone from your EOC get in touch with the various departments and check on their supply and equipment inventories. Know what your "par levels" are and for how long you can provide certain services without restocking. This is where the 'Joint Commission" "96 hour" rule comes into effect. Refer to your matrix (this should have been developed earlier) as a reference point for this information, in addition to the department reports you'll be receiving.  These are the type of tools developed earlier by you for your facility that will prove to be invaluable to you now during a disaster. As I stated earlier in the chapter, I would be using a hurricane response to illustrate some of the ways in which we will assist the community while responding to a disaster. While that was the case here, the principles remain the same for any disaster response by a healthcare facility in time of community need and crisis.

 Use what's written here as a guide and a starting point for you to develop your own facility response protocols. As we move forward, we strive not only for the facility to return to normal operations, but for our community to return to normal as well.

Response could last hours to even several days, again, to be determined by the type, severity and duration of the disaster. Think about the different devastating effects from Hurricanes, Tornadoes, Floods, Landslides, Earthquakes etc., to name just a few. Working alongside and with the community is part of our mandate. Being able to provide quality healthcare to our community members on a daily basis makes us an asset to that community. Being able to provide those services under duress and in times of crisis, make us a critical part of the community infrastructure.

In the next chapter we will look at some of the various things we will need to do in the "Recovery Phase", both for our facility and for our community.

# CHAPTER 5

## Recovery, Review & Moving Forward

Let's begin this chapter by discussing exactly what recovery means for a healthcare facility and when does it start. As far as when "Recovery" starts that's fairly easy to answer. It really starts almost immediately after our initial response begins. They'll actually run in tandem with each other. As far as what it is, in its simplest definition it means going back to "normal" or "pre-disaster" mode. I prefer to describe full recovery as the facilities ability to return to a "regular operating schedule". This would include things such as normal work schedules and time off for staff, returning rented or leased equipment, restocking depleted supplies and then a return to providing all featured facility services. The difference for some may just be semantics, but to me "normal" and "pre-disaster" are subjective. What is normal before a disaster may not be normal now. You may not be physically capable of returning the facility to pre-disaster condition due to damage received from a given disaster. The facility may be unusable in its present condition and or may be unable to resume its former use. If that is indeed the case, then you end up with a post disaster "new normal" for your facility. That's why I feel "Regular operating schedule" is a much more descriptive and accurate term in this case.

The other basic things we need to discuss concerning recovery are the different types of recovery, the major differences between them and how they overlap each other. For all intent and purposes, recovery, much like patient symptoms we encounter in healthcare facilities come two ways. For patients, we refer to them as acute (immediate or short term) and chronic (long term or extended term) symptoms. For "Recovery" it's *Short and Long Term Recovery*.

Immediate recovery is just that. The basics and essentials over the first two to three days. Short term recovery by its very nature will be things that are normally restored, repaired or replaced over a period of a week to a month. Long term recovery however, can and usually will take many weeks, months and may even take years in some cases to accomplish. A term I use when referring to recovery that often takes more than five years to realize is *"Extended Recovery"*.

I'd like to illustrate the points just made by utilizing the following example from the 9/11 terrorist attack on New York City. This will be by no means a thorough or detailed account of the event or their recovery effort. Rather, it will be a small snapshot as to what was done and over what time period it was completed. New York City dealt with the immediate issues at hand first, victims, survivors and safe guarding the crime scene while trying to maintain a safe environment for the first responders. Hospitals and healthcare facilities prepped for mass casualties that unfortunately never materialized. Food, housing, trying to restore electric power, water and communications in downtown Manhattan were the "Short Term Recovery" efforts they keyed on. From there, they eventually went on to the debris field, first going through everything, looking for survivors. After that, shifting through the debris field to recover bodies, body parts and personal items and then removing them (this process still continues some 15 years later with items still being found).

Over a long period of time, there was demolition of the partially destroyed buildings, opening of roadways and of subway and rail lines. Businesses were relocated, residents were also relocated. The economic impact alone was devastating for the area. Many businesses simply closed up. This went on for days, weeks, months and yes, even years.

This was part of their "Long Term Recovery". And while this was all going on, New York City started rebuilding lower Manhattan.

In 2014 after many, many years (13) and controversy, they finally finished "One World Trade Center", informally known as the "Freedom Tower". Its opened in late 2014. There is currently no start date for world Trade Center number Two. Number three World Trade Center is scheduled to be completed by 2016 and opened in 2018. Number four World Trade Center opened in November of 2013. And number seven World Trade Center was completed and opened in 2006. It currently is the only reconstructed building to be fully occupied. When all is said and done, it will be almost twenty years before the area in lower Manhattan is totally rebuilt and reopened.

This, along with the deleterious health effects experienced by the first responders would be an example of "Extended Recovery", which the city is still dealing with. And as mentioned earlier, the negative economic aspect is something that still lingers to this day for New York City. And while New York City never shut down during this disaster it does continue to heal and rebuild some fourteen years later. It will continue to do so. Lower Manhattan does not look like it did on September 10th or even September 11th, 2001.
But it has returned to a "New Normal" and "Regular Operating Schedule". Much the same can be said for places like Oklahoma City, the City of New Orleans and Joplin, Missouri. All have been through difficult times along with all of the various phases of recovery.

Now let's take a look at your structured recovery process for a healthcare facility impacted by a disaster. Continuing to use the hurricane scenario, what are some of the first things we are going to address as "Short Term Recovery" items? Obviously we'll want to look at our utilities right away. Did we lose regular power during the storm? Has it been restored, or are we still utilizing "Emergency Power" systems? These are the first questions you need to answer. If power is not an issue that's fine, but what about your water supply? Was it impacted or disrupted by the storm? If so, what do you need to implement (refer to your CEMP or EOP) so that you can continue to provide services? Remember, water is needed for drinking, washing, cleaning, cooling and sanitation. It is vital to a healthcare facility. Have you completed your facility damage assessment that we mentioned earlier? What kind of damage (if any) was sustained, minor (broken windows), moderate (trees down), severe (structural damage to buildings).

Something important that you'll want to review with your staff again at this time, is their inventory levels. Depending on the level of intensity and subsequent damage to the surrounding community, roads may be impassable which will make resupply difficult or non-existent at least over a short period on time. Have the department's report to you, given the present situation and with the anticipation of additional patients how long they feel they can operate without resupply. You'll then be able to develop a team strategy with Administration on how to proceed. This would also be the time to contact surrounding healthcare facilities that may not have been affected to see what supplies could be lent or borrowed (the importance of networking again) to your facility. Resupply of food, water, linens, medical supplies and pharmaceuticals are all items that should be on your list. As you start your recovery you'll also be going into a demobilization mode at the same time. As I mentioned earlier, things like portable lighting units or generators, heavy equipment, trucks and pumps are among some of the items you may have borrowed or leased and can now be returned.

A quick note here to save yourself some needless aggravation later. Whenever you lease, rent or borrow a piece of equipment, make sure you take pictures of it. Also take pictures of it again when you are returning it. This will help prevent any claims against you or your facility that the item was returned damaged (just like a car rental).

If the community was negatively impacted by the storm, which in all probability it was, Healthcare services will be one of the very first things the community will need restored. In addition to speaking to your departments about their "par levels" with supplies, have them also report to you on their staffing levels. How many staff members do they have working? Are they able to provide services to all three shifts? Have their work shift hours been extended, if so, are they being monitored and are they receiving proper rest periods when needed. Start thinking about returning your staff to their regular work hours and time off. It's important to note that as long as your staff remain viable, the facility will as well.

As far as the roads being impassable, you should assess what needs to be done. Then contact your local Public works Department through your EOC to the local EOC. See if priority can (should) be given to the access roads from the community to your facility. You're part of the Community Critical Infrastructure and as such will probably be given priority so you can continue to function in that role. Did you sustain any flooding in any of your buildings, and if so did it impact patient services in any way? If yes, what was the degree of flooding and how much damage was sustained as a result of it. Can you make that determination now by looking at it without making an in-depth damage assessment? Remember, water and the damage caused by it are tricky things to deal with. After everything is cleaned up it may look fine, but lurking behind the walls and under the floor tiles could be the beginning of a major mold issue. And if you end up with serious mold problems the health of your patients and staff could be at risk. Make sure that your post storm/flood building inspection takes all of these concerns in to account. I'd like to illustrate what I'm talking about by recounting the following story to you.

On our campus there was a separate building which had been built in the 1940's as a residence for the "Superintendent" (later C.E.O.) of the hospital. It was used for this purpose up until the middle 1990's. After that point for several years it was utilized for storage then set up as office space for several outside contractors who were performing various duties for the corporation. By 2004 it stood vacant. The building was equipped with an alarm system that was monitored by Hospital Security. The lock system was also "off-master" and Engineering and Maintenance needed to go through Security to gain access to check on the building systems. Somewhere along the line the building fell off the radar. As a result, it went through an entire winter without being checked internally. Come Spring the following year, Engineering had occasion to go into the building. What they found was a complete nightmare. The water pipes had frozen and burst which flooded the basement and first floor.

All of the wooden floors on the first floor were warped so bad they seemed like waves on the ocean. Worse than that was the half inch thick mold that covered all of the walls and ceiling. The building was sealed up and once it was determined that it was unusable, ended up being torn down. To reiterate my earlier words of caution, be very, very careful when it comes to sustained water damage. Now, what about broken windows, doors, building roofs, etc. Are you addressing them in a timely fashion? Can your engineering and maintenance staff handle what needs to be done for you?

Do you need outside help for these restoration services? I know these are a lot of questions being thrown at you. But you know yourself, having been through this very scenario or one similar to it that that's what it's all about. When you're going through a crisis situation, you're asking yourself these very same questions and you're also trying to formulate answers and responses at the same time. For those that have been through it, it's quite a juggling act. For those few of you that haven't, it really amounts to a "Baptism by Fire". Let's shift gears for a moment and go outside of the facility. What do your grounds look like after the storm? Are branches or trees down? Did this cause any property damage to buildings, vehicles or injuries to individuals? Do you have a proprietary grounds department that can handle the cleanup or is it a contracted service? If contracted, how soon can workers arrive on your campus and start working? Or, will they respond somewhere else first for more money? After storms, these type of service providers can pretty much name their own price. It's something you have to consider and plan accordingly for with some kind of "Plan B" should that happen.

While we're outside, so to speak, let's also take a look at a "Long Term" recovery item and that's structural building damage. Depending on the nature of it; wall cracks, foundation cracks, loss of roof, ruptured water, power, gas lines, will be major items to deal with.

The time frame for recovery and the assistance needed, will also be a major undertaking. If the damaged building is occupied, you'll first need to determine if it is safe for individuals to remain there. You'll need technical professional individuals to make this type of determination, people like; architects, structural or civil engineers, building inspectors, electrical/water/gas inspectors and the local Fire Marshal should all be considered and brought in as necessary.

Again, go through your facility EOC to the local EOC for this specialized technical assistance. If there is any doubt whatsoever or if there will be a delay in getting the proper people on scene to make a determination, move the people and/or patients as soon as possible. Do not risk their safety for even one moment. Consider these people your family and respond accordingly. This then leads us to another concern. Once the building or facility has been inspected by trained technical professionals, what if it is determined that it cannot be used again for its original purpose. How should you proceed? Although the decision will certainly not be yours alone to make, you will still have important input into the entire process. If it is no longer acceptable say, as a patient area, perhaps it can be converted into office space, a meeting area, staging area or even storage. This particular function will overlap between long term and extended recovery. In extreme cases, where the damage is too severe or too costly to repair, the building may have to be closed permanently and possibly even demolished.

While you're evaluating the recovery of your main facility, you'll also want to be checking on the status of any of your off-site "alternate care sites". What condition are they in? Is there any physical damage to them? If they were activated are they still functioning or did they have to close down for some reason? Can you deactivate them now or are they still needed as part of your overall recovery process? And what about restocking of necessary items? Do you have the capacity or ability to accomplish this?

I think at this point, if it wasn't totally clear before, it should be now regarding the different phases of *"Recovery"* and many of the different aspects that have an impact on it. Not the least of which is funding. And funding is directly tied to and affected by documentation. It's all just really one big continuous cycle.

One last important question; When do you know "Recovery" is complete? Unfortunately, there is no clear cut answer to that question. There may not be a clear line of delineation to make that determination. As we've pointed out "Recovery" can stretch out over years. In that case, you need to make a declaration affirming that the "Recovery" phase and effort is at an end. Once you are able to return to "normal operating procedures" that declaration could be made. It's really subjective though, given the fact, that each facility is different as well as each disaster's impact. I'm going to finish up on recovery by listing some bullet items to be aware of for each phase. This is not a comprehensive list but simply an overview for illustrative purposes.

IMMEDIATE
Critical Infrastructure -Water, power, HVAC

SHORT TERM
Resupply - food, medical supplies
Roadways
Minor repairs
Staffing
Site evaluation, operations

LONG TERM
Major structural repairs
Rebuilding
Mitigation measures
Medical monitoring of staff & first responders

EXTENDED
Site re-designation
Major new construction, demolition or repurposing
Long term medical and mental health monitoring
Economic reimbursement and financial recovery

I'm certain anyone of you can think of at least another dozen or so items to add to this or your own list. As long as you're thinking about it, you're still ahead of the game. There's nothing worse than trying to play catch-up in Healthcare Emergency Management, especially during a crisis.

Let's move on to the *"Review"* portion of this chapter. Review and by its very nature "Evaluation" are some of the most important parts of the overall emergency management cycle. This is actually the learning and educational portion not only for staff, but for us "The Emergency Manager" as well. In part, it gives us the opportunity to look at what we've written as plans and see how they fared, during the disaster response. Review also needs to be ongoing and constant if it's going to be an effective tool. As I've mentioned earlier, review and evaluation should be done after every single drill, exercise or actual response.

There are many different forms of review, so let's go over a few of them now. First of all, either during or right after a scheduled event or a real time event you have what's referred to as an *"initial debriefing"*. For our purposes, it refers to the immediate information received from participants in an exercise or drill or from our responders during an actual event. It is normally verbal and not written down. This will be raw data and can be used to adapt our responses accordingly during these events.

The next term or part in the review process is a *"Hotwash"*. This term itself, comes to us from the U.S. Army and troops in the field. The main purpose of a hotwash session is to identify strengths and weaknesses of the response to a given event, staged or real time. As a result, *"Hotwash"* is a term that has been utilized in recent years by Emergency Management Managers and responders alike.

It is normally utilized as a tool for an after disaster briefing for all of the individuals, departments and agencies involved to review what worked well, identify what didn't, and then to analyze what needs to be accomplished for necessary improvements. Depending on how far you plan to take it, it may also identify what group or agency will be responsible for the identified improvements, and then what kind of timeline will be needed for those improvements.

The other term used for this purpose and now seems to be interchangeable is the term *"critique"*. A critique is a "detailed analysis and assessment" of something, in this case, our response to a scheduled or real time event.  Some Emergency Managers still make a differentiation in the two, citing that a "Critique" is the more detailed of the two. In addition, many times "critiques" will be set up in a meeting format. For us in healthcare it may be the Emergency Management Committee or the Safety Committee that performs this function. Either way, it's a very good idea to utilize this format and allow others to have both insight and input into the entire process. I'll let you decide for yourself, whether you want to use one or both of them. Remember, what works best for you, your program and facility is the "right way" for that place and time.

This now brings us to a phrase known as *"lessons learned,"* which after reviewing our analyzed data, is intended to guide all of our future responses. The main focus of this, is in avoiding errors made in previous responses, and then keeping us from repeating them in future ones.

All of the above now bring us to the items that are used to create an *"After Action Report"* (AAR). In some parts of the country this is referred to as an "After Action Review", regardless, it is the same thing and serves the same purpose. For clarity, we'll refer to it simply as an AAR. An AAR is a structured and detailed review for analyzing *what* happened, *why* it happened, and *how* it can be done better by the participants and those who responded to event or disaster. This takes the critique and hotwash to the next level and beyond. AAR's in the formal sense were also originally developed by the U.S. Army just like the hotwash. Informal debriefings however, have existed in some way, shape, or form almost forever.

There really is nothing that difficult in conducting a proper AAR for your purposes. Just like what you did in setting up a group to write your EOP or HVA you'll do here as well. Probably the most difficult thing will be establishing a meeting date and time that will be beneficial to everyone.

Before you have everyone attend, set up an agenda of what you wish to accomplish, send it out to everyone invited and then stick to it on the meeting day. If you do this, it will make your life much easier and hopefully will keep everyone on point. If other items come up that are not on the agenda and they are not urgent, table them for another meeting if possible. If the right people are present, they will realize that this is an on-going process and should not only be receptive to the process but be active participants in it as well.

Something else to keep in mind regarding AAR's, is that the more often you complete them the more proficient at preparing them you'll become. This will also bring up the quality of your reports as you'll have a comfort level, along with other staff in preparing them. To that end, I would strongly suggest that you complete an AAR after almost everything (event) that occurs in regard to Emergency Management at your facility. This means every tabletop, functional and full scale exercise, after every drill, even after any and all EOC activations.

Document everything using an established format. Do not take short cuts, it will only hurt you later on. Keep this documentation, along with all of your documentation in your accreditation binder and electronically. Accreditation Surveyors can't help but look favorably on this type of attention to detail and thoroughness on your part. This will help let them know you're a professional and that you take pride in your work.

And while we're on the subject, if you have the opportunity to speak about one of your drills or exercises make sure you speak to the AAR as well. As long as you've followed up and document it, don't be afraid to point out some of the shortcomings you've found. Let them know what you did to resolve or correct the identified issue. I've seen too many practitioners hesitant to air what they consider their "dirty laundry".

The surveyors know that no facility is perfect, so don't fool yourself into believing that your facility has no issues. Part of your job is in identifying those weaknesses and then working with the appropriate staff to improve and overcome them. They'll recognize and appreciate your effort more, if you show that you've truly evaluated, modified and updated your response process after a drill, exercise or actual event.

Having gone over the review process, I'd like to speak a little bit about the evaluation portion which goes hand in hand with review. I won't spend a lot of time on this as it should be obvious at this point that this is something that has to be done for every event. It's another one of those "tools" that we have at our disposal to perform our job and make our facility's safer places for our patients and staff.

One of the formal definitions of *"Evaluate"* is; to judge or determine the significance, worth, or quality of something. When applied to the Emergency Management process for Healthcare Facilities, we're looking at our plans, procedures and responses to events. We need to stay focused and be truly objective when evaluating or the entire process loses its effectiveness and integrity.

As I mentioned earlier, glossing over shortcomings, leaving out data that might be detrimental, or manipulating statistics and data to make the facility look good, is counterproductive and unprofessional. If we don't learn from our errors, omissions and mistakes we are bound to repeat them. If at all possible, utilize other individuals in the evaluation process. This will make the process easier for you and provide a fresh set of "eyes" on the material at hand. It may also provide a wake-up call for you, since the individual will be looking at it from an entirely different angle and slant. You might actually pick up on something that you did not initially see yourself. When your evaluation has been completed, bring it before the Emergency Management Committee for review and comment. After that's been completed make sure Administration is appraised and then you can add it to your documentation cache.

I'd like to finish up this chapter by discussing *"follow-up and moving forward"* and how it fits into the whole process. We've already spoken a little bit about this when we discussed our "After Action Reports".

This then, is what we are working toward every time we respond to a drill, exercise or real life event. It's with the ultimate knowledge that we can always do better. No matter what we've done in the past, once our evaluations are completed, there will always be room for improvement.

Again, this is part of our on-going education process that we spoke about earlier. It also allows us to adapt and change as the situation or needs warrant it. We then take what was learned in the AAR's and critiques and "tweak" or "modify" our procedures, our plans and finally our program. This should allow us to provide a better response for the next event. If it doesn't bring the desired effect, then we need to revisit it again and change it. This process never ends. It simply moves from response to response. That's why earlier we said that the CEMP and EOP are pretty much "living" documents that are never truly finished. Circumstances are never the same, so all future responses will vary, even for the same type of an event. Sometimes the best you can hope for in a situation is to NOT repeat past mistakes or errors. If you do, then you haven't learned anything and you may end up putting yourself, your staff and your facility at further risk.

When performing *"following-up"* you also want to apply it to your AAR's. This becomes extremely important during the accreditation surveys of your facility. One of the things they will pick up on is items that have been identified by you as needing improvement. They will look at and ask you for documentation showing the improvement or resolution of the identified problem. They will also want to make sure that if the issue was resolved, it didn't appear again. Make sure to close the loop on these items. Remember, this is part of what we're there for.

And finally one more word on *""follow-up"*, this time concerning your Emergency Management Committee. The same thing that we just spoke about regarding your AAR's applies just as much to your Emergency Management Committee's meeting minutes and agenda.

Whatever you do, do not leave items hanging out there month after month without action or resolution. Do what you can to make an improvement or correction and then remove it from your agenda items. If you find out it is something that you will not be able to address, then at least forward that information along with any recommendations to Administration or the Hospital Safety Committee to show that you took some sort of proactive, appropriate action. This doesn't necessarily resolve the issue identified, but it does show that you did address it and it gets it off your agenda.

Once you've prepared, mitigated, responded, critiqued, evaluated and followed up an event, you need to now look at *"moving forward"*. At this point in time, you may still be involved in the recovery process. Recall how we've spoken about the different phases of recovery; *immediate, short term, long term and extended recovery.* Think of New Orleans and New York City as just two places where recovery is still going on ten and fifteen years later. In New York City, some of their buildings aren't slated to be completed until the year 2020 and possibly beyond. That's a twenty year timeframe we're possibly looking at. Now is the time to see if major changes are needed at your facility or in your program and plans. What actions will prepare you better for the next time? How can you make your facility and staff more secure for the future? "Moving Forward" is as much a "state of mind" as it is part of the practical process of Emergency Management.

 It means recognizing that a disaster has occurred and that the facility responded to it. This may seem elementary, but in cases where the community has lost many lives as a result of the disaster, moving forward is not always easy. These fatalities may have been family or friends and not strangers. Think of the Newtown school shooting in Connecticut.

The post trauma of that incident affected almost everyone in that community including the Police, EMS and healthcare professionals. As a result of this, moving forward then becomes as much of a "healing process" as a post recovery one. We as human beings, need to wind down from the disaster mode as much as our facility does, maybe even more. There is a limit that everyone has when working in this type of environment or situation.

If pushed too far and for too long, we run the risk of adding to the victim list. People have a breaking point and can "burnout" if left too long in crisis mode. Be aware of that, and try your best to prevent it. Rather, moving forward brings the facility and staff back into a more normal operational mode. We do well with emergencies, but we do even better when we get through them to the other side and back to normal. It gives us balance and allows us the ability to recharge our emotional batteries. It also gives us hope that things will get better. Don't ever underestimate or shortchange the importance of hope. Many a life has been saved based solely on an individual's hope and will to survive.

Moving forward also gives us the ability to look back and reflect, not only on our response to a disaster and our effectiveness, but how it impacted our facility and ourselves as well. Surviving a disaster is one thing, coming through it a better individual is another. Not everyone will. Not everything will be the same. Not everything will be better.

Don't be afraid to praise and encourage staff for their actions during the crisis. A little reassurance from you as the Emergency Manager can go a long way in providing some comfort to them.

This small act of kindness helps allow them to continue in their mission. And while you're at it, don't forget to give yourself a little pat on the back as well. Having been in your shoes, I'm sure you've earned and deserve it as well.

# Part II

# When the event strikes you

## CHAPTER 6

## Preparing for the "Big One"

In part I of this book we concentrated on what our role and response was to a Community based disaster. We looked at things like critical infrastructure and our role as a community asset and how we could continue to provide services in this great time of need.

In part II, we'll take those concepts to entirely new level and different direction as we consider how to provide that role when the facility itself is the site of the disaster. We'll look at things like partnerships, resources, training and networking and how they can benefit us during a crisis. We'll discuss the role of our "First Responders" and how we can take preventative measures so that they do not become secondary victims when responding. We'll also discuss the concept of running parallel EOC's (yes, I know this flies in the face of convention) and how this could benefit you in both response and operations. We're going to explore various scenarios such as; explosions, hurricanes, tornadoes, major floods, active shooters, fires and then look at some real life examples to see how they affected specific hospitals and in some cases devastated them. We'll review what mitigation and preventative measures they took and whether or not those were effective. We'll also take a look at the degree of damage that each facility sustained, and see if they were able to cope with and overcome it or if indeed, it overwhelmed them.

We'll then discuss any commonalities that we may find in reviewing these different case histories and see what lessons, if any, can be learned from them.

Remember, we may make mistakes, we just do not want to repeat any previous ones. Alright, once you've completed your "Hazard Vulnerability Analysis" (HVA), you've run your drills and exercises and you have completed your evaluations and "After Action Reports". You're now ready for anything, right? Well, maybe, maybe not. As a Healthcare Emergency Manager, besides education, experience and training, you need to also rely on your instincts and gut feelings. In the back of your mind or in the pit of your stomach, what keeps you up at night when you worry about your facility and its ability to function if a particular disaster struck it directly? What about the unknown or the unexpected? How do you prepare for something that you don't know will happen? For one thing, having a program in place that doesn't rely overly on specific responses to disasters but rather the ability of your facility to respond in an organized and effective manner, regardless of the event that occurs at or around it can make a big difference. And as you'll see, that can only be accomplished through effective, comprehensive preparation and training by you, the Healthcare Emergency Manager.

To begin that process, you'll need to refer back to your HVA and focus most of your energies on at least the top three (if not the top five) identified hazards. These will be the ones that have the highest ratings and the ones that you'll prioritize and concentrate your energies. Again, these will be different, around the country, based upon such things as climate, geographical location, type and size of facility and even population density. But let's begin simply by looking at fires that can occur within our facility.

The threat of fire is something that we all dread and deal with on a daily basis. Fires occurring anywhere are frightening things but a fire in a healthcare facility can be devastating. We all know the potential catastrophic results that can occur from a large or uncontrolled fire within our facilities. This is really something that needs to be analyzed by you as the Healthcare Emergency Manager.

Think of the various areas within your facility, and the issue or difficulties they may present. What additional precautions have you taken or training have you dedicated to these areas? For instance, a fire in any of your lab areas or even in the engineering areas present the additional issue of chemicals and hazardous materials being entered into the mix. How about your pediatric unit, where is it located within your facility? A fire in this area could have dire consequences if the staff don't respond properly and control the patients (children's) anxiety and fear. How about your intensive care units (ICU'S)? Or your Maternity unit. Its labor intensive (pardon the pun) to move these kind of patients, beds and equipment.

Do you have secure or open psychiatric facilities? What about the challenges that evacuating this type of patient population poses. And these are only some of the patient areas to be concerned about. What about your staff, visitors, vendors? They all represent unique issues that you must find the solutions to, prior to an event in order to protect them . These will be some of things that we look at in the next chapter. But for now, back to fires. Are you truly comfortable with your facility's response to "fire drills" and or "evacuation drills"? Does your staff take them seriously enough or do they go through them by rote. Do they seem apathetic to the whole process?

I have found that in my experience, these types of drills are some of the hardest to get staff to take seriously and actually become involved in.

It's a little like when people fly commercial airlines, and the flight attendant is going over the safety precautions and instructions. Very few people pay attention and even fewer of them listen to and follow along. Yet these instructions are critical and could save the passengers life in an emergency. The same is true of fire and evacuation drills. And in our case it's not just the staff member but patients' lives as well.

How often do you conduct these types of drills? Do you perform the minimum required by code or standard or do you go above and beyond? How do you quantify the effectiveness of the outcomes of these and your program regarding the drills? What about training? Is there enough, who puts it together, who reviews it and who conducts it? These are all basic questions that you need to ask and have answers too.

I'd like to recount two separate fire incidents we had at our facility to illustrate the importance of training and staff response. In the thirty-five years I was there, we had many fires, mostly small in nature, producing minimal damage. We also had several large ones that caused significant damage to the facility. The two that I would like to speak about, involved fires in vastly different patient areas (our ICU and a secured psychiatric unit) and required patient and staff evacuation from the affected units.

First of all, I'd like to speak about a fire which took place on our fourteenth floor, which was an in-patient secure psychiatric unit. While smoking was prohibited at the time and patients were not allowed matches or lighters, a patient still managed to start a fire in their room.

It was later determined that a visitor had passed the patient a book of matches during visiting hours. To complicate things even further, this fire took place at 2:30 in the morning.

Because the unit is a secured psychiatric unit, all fire extinguishers and fire alarm pull stations are secured as well. They must be psychically unlocked by staff to activate or use them. The patient's bed which he had set afire, was almost fully engulfed by the time staff discovered and responded to it. They safely removed the patients from that room and began an entire unit evacuation. Several staff members also began to use several ABC fire extinguishers on the fire, which were on the unit.

In a total of sixteen minutes, staff members were able to safely evacuate fifty-two (52) patients off the unit and down a stairwell to the first floor auditorium. The reason staff took the stairwell and not the elevator was so that they could keep track of all the patients and keep them together. Staff performed headcounts when they left the unit and when they arrived at the auditorium.

They also repeated this process when leaving the auditorium and returning to the floor. Elevators were utilized for the trip back up to the unit as there were now additional staff to track and safeguard the patients. Fire damage was confined to one side of the room, requiring it to be taken out of service. However, smoke damage affected the room and the entire unit as did the powder from the extinguishers that were used. It took Housekeeping staff approximately two hours to return the unit to habitable status. No patients or staff were injured as a result of this fire. The quick response and actions of the staff were directly cited for the success of containing both the fire and safely evacuating patients and staff from the unit.

The reason they did so well was a direct result of the extra time and effort that was spent in training all of the psychiatric staff. Because of the uniqueness of their unit, they need to react instantaneously when a crisis arises. The staff did just that. We were very fortunate and thankful for the results of that early morning response.

The second fire of note that I'd like to speak about that occurred at our facility, was one that broke out in our "Intensive Care Unit" (ICU). Although smoking was not allowed in our facility at the time, the patient (female) had secreted a cigarette and matches inside of her underwear. As a result of her illness she had been placed in an isolation room and was on oxygen. During the early morning hours she attempted to light her cigarette within this area.

With the oxygen flowing, she managed to ignite the bed linens and oxygen tubing instead. The nurse, who was sitting at the nurse's station almost directly across from this room, immediately noticed the flash of flame and sounded the fire alarm. At the time of the incident, our ICU's (Medical, Surgical, and Cardiac) were separated across the hall in the same common area from each other. Thus, the fire in one area would impact all patients in all of the ICU areas. The fire broke out at approximately 7:00 am in the morning.

This, at the time, was shift change for the clinical staff. There were a total of twenty-seven patients in the three ICU areas. In seven minutes, from the time of the alarm, all twenty-seven patients had been moved safely to other areas. The patient who had set the fire, received minor burns and was the only individual injured. Because it was shift change, there were additional clinical staff present to assist in the evacuation.

In addition, Engineering staff, Security staff, Food service workers and housekeepers were instrumental is assisting the clinical staff with moving the patients, their beds and their equipment. In less than twelve hours, the area had been vented, equipment repaired, and the room repainted.

All of the patients were returned to their areas within that day. This is just another example of how good training pays off. Plus, the realization that people react differently when they know the situation is real and that lives (theirs as well as others) may depend on them taking immediate action. The fact is, that all employees working in the area that day and not just clinical staff, made the difference in preventing a potential catastrophe. I sincerely hope that if you or your facility ever encounters a similar situation, that the results are as positive as mine were.

On the other side of the coin, I'd like to recount two different instances wherein some employees responded emotionally rather than logically to a perceived crisis. The first one occurred in the autumn of 2001, after September 11th. Unexpectedly, on a beautiful day, the hospital lost all power. There was no apparent reason for it. It was later determined that a fire at the Power Utility sub-station caused the outage. What was worse, was that the emergency power generators did not kick in immediately. People started to panic, and rumors spread of another terrorist attack.

You have to remember that everyone's nerves were still emotionally raw from the events on 9/11. At this point, many staff members started leaving their units and the building.

Even after the emergency power had come on, staff still continued to leave the building for up to thirty minutes (30) after the initial power outage. The Public Safety Department eventually had to send out Officers to explain the situation to staff and advise them that they needed to return to their units. Eventually, almost everyone did. While not one of the finer moments for us at our facility, it simply illustrates that even with good training, sometimes emotions are too strong and take over in place of common sense and logical thinking. In light of the events that preceded these actions, I don't really think any amount or type of training would have made a difference on this particular day.

The second incident, with similar consequences and actions was completely different in nature from the previous one. It was also separated from the earlier incident by a period of over ten years. The second incident occurred in August of 2012. While it can't be compared with what California has to deal with, New York does also have earthquake faults running through the State.

It's generally not something we think too much about or spend much time preparing for. It's certainly not an item that makes our top three on our Hazard Vulnerability Analysis (HVA). However, having stated that, we did experience an earthquake strong enough to shake our twenty (20) story tower. Vibrations were felt throughout. People on the upper floors felt it even stronger. At this point, again, emotions took over and panic started. Staff and others started leaving their units and the building. Some rumors even had that a plane had hit the hospital similar to September 11th. In a "Déjà vu" episode, once it was determined what had happened, and that the building was safe, Public Safety Officers again had to go around requesting staff return to their units.

On the positive side with both incidents, patient care was not adversely effected due to the fact that many staff did stay at their posts and tended to their patients. Healthcare employees tend to be a different breed of individual and look to their respective professions more a calling as opposed to just a job. We are fortunate indeed to work among these people in this industry.

What I've tried to illustrate here and earlier in the book is that "preparing for the big one", like anything else, requires a lot of practice and training. We spoke about "Rick Resorla" and the difference his training program made with his staff on that fateful day on September 11th, 2001. We may never be faced with that type of catastrophic incident, but the need for good training is just as important to you and your facility, whether it's the everyday trials and tribulations we deal with or an actual disaster. Training, more often than not, is the one thing that will get you through a crisis. You can have all of the experts in the world, but if they can't communicate or put their knowledge to practical use, if they haven't been trained, those experts and that knowledge is negated.

Therefore, the basis and secret of good facility preparation, regardless of the disaster or hazard identified, is having a thorough, well rounded and multi-faceted training program. One that's adaptable when times and situations change.

While we're on the subject of preparation and training, I'm not just referring to your facility staff. I'm referring also to the individuals who come to or respond to your facility for various reasons. Groups like your local or State Office of Emergency Management (OEM), EMS, the local Fire Department, the Police or Sherriff's Department, or County Fire Marshal's.

All of them need to receive and participate in training at your facility. And don't forget about your vendors as well. Remember, these are all community partners for you and they also have a stake in maintaining the operations of your facility on a daily basis and in times of crisis. Don't forget about them, reach out to them and see what you can provide to them to make their jobs easier when they respond to your facility.

As mentioned at the beginning of this chapter, utilizing your HVA and focusing on the top 3-5 identified hazards is a good place to start this process. Review the items you've chosen, and then break them down by analyzing what you would need and who would be the most likely individuals or agencies to respond to your facility for that particular crisis. Think about past events and disaster responses in your community or at your facility. Who showed up? Who did you reach out to? How did you interact with these agencies? Did you personally know any of the responders? By thinking and planning ahead, you can then begin to plan and prepare your training and exercises around that focal point. Remember, these responders and the agencies they work for are just as valuable a resource to you as your utilities and supplies.

Keeping that in mind, we'll be looking at events that occur at our facility and directly impact in our ability to operate, such as; explosions, active shooters, devastating natural disasters.

We'll look at ways to create working partnerships with all of the parties involved in our process. We'll then review the needs not only of our own facility and its populations, but also of the people and agencies that respond to our facility in time of crisis. This is where the planning, preparation and exercise process takes on even greater importance.

As you'll see in the next chapter, we will find ways of enhancing our response by utilizing our responders as an extension of our entire Emergency Management Program. Think of the team process or if you will, the "bigger picture" when it comes to resources and response. How do we incorporate all of the players as integral members of our team?

As we move forward into the next chapter, we'll take this knowledge and put it into practical use by looking at ways to maintain our facility if at all possible, during a disaster. We'll also focus on just what we need to do to protect our patients, staff, vendors, visitors and first responders during that disaster. What mitigation measures can we institute at our facility that will allow our first responders to safely respond to our facility when we need them? Again, as stated earlier, the last thing we want is for the people who are coming to assist us, to become victims themselves.

# CHAPTER 7

## Protecting staff, patients & first responders

Having left off in the last chapter, one of the things that we've stressed throughout this book is the importance of the training that we provide. But when we're looking at protecting staff, we can't just focus on "book learning" or even how many times a staff member receives training during a given year or other time frame. There are many other things we need to consider and discuss that will make that training truly effective. Repetition is one of those ways to assure positive results. In this way, when a crisis arises or a disaster event occurs, the training that the staff received will "kick in" and carry them through the event uninjured. When staff have truly been trained properly, they don't consciously think about that training or what they need to do in a crisis situation. It becomes second nature, they perform it instinctually if you will.

Let me give you a short example of what I'm referring to. My wife who is a nurse and who was employed at the same hospital as myself was certified in CPR. I was as well, and at that time was also a certified instructor in CPR. It was a weekend during the summer and my neighbors were having an outdoor party. My neighbor's mother was living with them at the time in an apartment attached to their house. With our windows and doors open, I heard my neighbor scream my wife's name and call for help. My wife immediately ran up from the basement and went out the door to our neighbors.

As she did, I automatically called 911 and requested an ambulance. Not quite sure what the issue was, I simply said it was a medical emergency.

Then, I too, ran next door to my neighbors. As I went into the apartment, I saw my neighbor kneeling on the floor next to her mother. Additionally, my wife was there, as were several other people. My wife had already checked for the pulse and ascertained that the woman was not breathing. We simply looked at each other and I asked "do you want breaths or compressions" (which at the time was the standard procedure). She said I'll start with compressions. At that time we began the CPR cycle on the woman. On the third cycle she started breathing and she regained her pulse. Our actions along with the quick response of EMS and a timely transport to the hospital were all part of the reason she survived that day, and thankfully many years after. While both my wife and I had been trained, we had never trained or practiced together, yet instinctively we were able to respond quickly and work efficiently as a team. This wasn't because we were extra special, but rather because the training we received had been done effectively. And when it was needed for this situation it "kicked in" automatically.

That is the type of training, regardless of what it is, that is needed to keep your staff safe and secure both during emergencies and everyday operations. We're not just talking about CPR, or needle stick training, or patient restraints, or fire response, or, for that matter "Disaster Response". To truly protect staff, this needs to be the mindset of the entire facility and should be the focus of all of the people providing that training.

Whether they're from Human Resources, Nursing Education, Central Training, or anywhere else, the entire facility training program should in some way overlap these basic principles. Training should never be viewed as a "necessary evil". Unfortunately, I have to say, that many all too many times that is indeed the case. We as trainers and instructors should never just "go through the motions" when performing our duties or giving training sessions. Whether it's new employee orientation, or annual refresher training, or even training for a new position, the training needs to be interactive and it has to be relevant to the employee. They need to actually be part of the training process and not just an end product of it.

Think of the fire safety training that is provided at your facility. What does it entail? Is it simply "book learning"? Have any of your staff ever actually handled a fire extinguisher? Do you have the ability to utilize the local Fire Academy for training staff? Have you ever looked into it? What about the local Fire Department or the local Fire Marshal's Office? Any one of these agencies may be able to provide "hands on" training or provide a trainer to come to your site to perform these functions. For instance, if you've never used a $CO_2$ (carbon dioxide) extinguisher it can scare the heck out of you when activated. Hands on training takes the unknown away. We at our facility, had the good fortune for many years of having the ability to provide staff training at our local "Fire Service Academy". This made a tremendous difference in our ability to quickly respond and contain small fires (Especially at the time when smoking was still permitted in hospitals). What about "evacuation drills"? Have you ever actually made staff physically leave an area or unit? Did they know they were being observed and timed?

Let them see and feel what it's like to perform this task. We once ran an evacuation exercise where we timed taking twenty-five "victims" (staff members) down the stairwell of out of the tower. Staff members who were utilized, used crutches, canes and in some cases were even blindfolded while being led out by other staff members. This gave us an entirely different perspective, on our ability, and the time it would take if we ever had to accomplish this task for real. What about HAZMAT response? Does your facility do this? Do you have a specific group of staff as a response team? How are they trained?

Have you taught them how to "don and doff their PPE" in the proper manner so that they don't cross contaminate? Have they ever had to suit up while being observed and timed? Have they done it outside in the heat or cold? Do they know what it feels like to be wearing a respirator to breathe? And what about respirator use training? You don't want to be surprised when an actual staff response is needed only to find out that the staff member who responded is claustrophobic and can't wear the mask and respirator?

Now what do you do? The same thing goes for disaster training, get the staff involved. Use moulage patients if you can, and have staff interact along every part of the way during the entire response process. This keeps your training program fresh and interesting. This will hopefully make your staff look forward to the training rather than dreading it.

These are the type of things I'm speaking about when I say to make the training interactive for your staff. It really makes all the difference in the world in what's learned and whether or not it's actually effective.

It's very easy to lose "adult learners" during a classroom or training session. We've all seen the "deer in the headlight look". We can't afford to do that, not if we want to get them home safely to their families at the end of the day.

I honestly wish I had all of the answers for you, in regard to how to actually provide a truly effective staff training program, so that your staff are protected, but I don't. No one does. Not every part of my program was equally effective. Sometimes the effectiveness of the training varied from year to year or from subject to subject. It was always a struggle to make it relevant, interesting and informative. I will tell you that perseverance and commitment on your part, as the Emergency Manager does make a difference and might be the one constant in all of this.

You have to "fight the good fight", you need to push the appropriate "higher ups" to get the proper training for your staff as well as for yourself. You won't always be successful in your attempts, but you do need to consistently keep trying. To digress a little, do you think that if Rick Rescorla had not been as persistent as he was in his training needs and requirements, that the amount of people that survived from his company during the September 11th would have been as high as it was? Remember, it wasn't for a lot of other companies. Just keep trying no matter what. You owe that much to yourself, your facility and certainly to your staff.

We've all dealt with hard headed administrators who only look at the "bottom dollar" (but remember, they have to answer to someone also). We need to show them that this will actually improve the bottom dollar in many ways. First, we'll be in compliance with accrediting agencies which means no bad press, fines or bad grades.

Second, with properly trained staff, in times of crisis, those staff members will still be able to perform their duties. This then means the facility can continue to operate. And if the facility continues to operate, it continues to generate revenue. So, you could argue logically, that all training benefits the bottom dollar of the facility. This is true both during times of crisis and during daily operations.

We should also remind them, that the facility's staff are really their greatest asset. And as such, it's an asset that should be trained and protected. Even this won't always work, but it is a valid and sound argument you can provide to make your case.

Let's take a little time and look at the ways we can and do protect out patient population. Patient safety comes in many, many forms in the healthcare industry. The majority of it is covered by the clinical staff and they do a great job of it. Simplifying it; doctors cover operative procedures, nurses cover infection control and treatment, pharmacists cover medications, and therapists cover rehabilitation. And these are just some of the individuals who interact with our patients. Housekeeping, food service, medical electronics and even engineering staff, on occasion, all have the need to directly interact with patients. Again, they all play an important part in protecting and safeguarding our patients.

So what exactly do we add to the mix? Well, for one thing, we provide, monitor and maintain the "Life Safety" systems throughout our facility. Things like our exit lights, emergency lights, fire alarms, pull stations, sprinkler systems, and fire extinguishers all play a part in emergency management. Do you still use "standpipe" systems? Are they maintained properly?

These, along with evacuation route maps placed throughout your facility provide a great deal of patient protection. Another way is making sure that all exit, fire and smoke doors are functioning properly and are not being blocked or locked. This applies to all patient unit hallways as well.

We all know what a constant problem this can be; linen carts, medication carts, portable medical equipment etc. always cluttering up the hallways. They need to be kept clear to provide a free "means of egress" for the patients and staff in the event of a unit evacuation for whatever reason. This should be a simple task, but in my experience, it's one of the most difficult things to keep in compliance with.

We certainly can't rely on staff to remove these items while assisting patient's in an emergency. It's just not feasible. They're responsible for too many other things. It's just one of those things that you need to monitor and educate on a continual basis.

Is your facility a "smoke free" facility? Nowadays, almost all facilities are. It's another way in which we reduce the likelihood of a fire at our facility, thus providing even more safety and protection to our patients.

What about patient education? Do you have "patient educators" at your facility? If so, what is covered under their program? Can you insert any information about the fire alarm activation or disaster notification? Something as simple as instructing them not to panic and to follow all clinical staff instructions will certainly help in your goal to provide a safe environment to your patients. What about a hospital broadcast channel? Many facilities have this type of system set up to disseminate information to patients over their television sets.

Do you have a "workplace violence prevention program" at your facility? This is something that protects both your facility's patients and staff as well.

Do you have a way of knowing if patients or staff have "orders of protection" for themselves? If so, do you or your Public Safety Department keep copies for your records? This is extremely important should an adverse event occur at your facility.

Explore online for up to date information and statistics regarding workplace violence. You may also want to enlist the aid of an outside expert or training company. One that I am aware of that specializes in this is, *"Nater Associates"* a nationally recognized and well respected company. Feedback on them from my colleagues has all been positive. But the choice is yours. As long as you are actively involved in the process, and are trying to provide as safe an environment as possible for your staff and patients, you'll be headed in the right direction.

Something else that ties in directly with workplace violence prevention is a phenomena that unfortunately, we are all too familiar with nowadays, and that is an *"Active Shooter"*.

Hospitals and healthcare facilities are no longer immune from these types of violent acts. You need to have a facility policy and some kind of working procedure in the event of an occurrence. To do this, you should definitely seek out the expertise, guidance and advise of your local law enforcement agency as they will be the ones responding to your call for help. This is a great opportunity to share information about your facility with them. Then, if they do respond, they will have a better idea of the internal make-up of your facility.

Ask them about potential staff training for your facility. If this is something they are unable to provide, then seek out an expert in the field (see above).

Once you have this in place, it would be a good idea to run and internal (facility) drill and then an exercise (with law enforcement) to educate your staff. Just don't make it too realistic. The purpose is to educate, not frighten staff.

Many facilities around the country post and distribute a "patient bill of rights" to all patients upon admission. We had this at our facility. In addition to this, our facility also had a "patient's bill of responsibilities". Among these were the fact that patients were required to follow all hospital regulations and follow all lawful instructions from staff in time of emergency. And while this didn't always occur, it was a further attempt on our part to provide a safe environment for our patients.

Again, these are just some of the many ways we provide protection to our patient population. I'm sure you've come up with some of your own ways as well. As I've stated several times in this book, the right way is what works best for you and at your facility. Don't ever be afraid to try something new or different. If it doesn't work out, change it around till it does.

Now we're going to explore some of the many ways in which we can provide a safe environment to the very people who will be coming to our facility to assist us in our time of need; "Our First Responders". Let's start first with our local Fire Department and their contingent of "Firefighters". Here on Long Island, New York, ninety-nine percent of the Community Fire Departments are staffed by volunteers.

These are people are just like you and I, our neighbors if you will, who are giving of their time and energy to make their communities safer places to live. Part of our jobs as healthcare Emergency Managers is to protect those protectors.

Now what are some of the things we can do to ensure their safety when responding to our campus and facility? The first thing we should do is get to know them. We do that by setting up face to face meetings. In smaller communities, you may actually get to know all of the department members. In larger ones, like the one I was part of, make sure you at least know who the Chiefs or Officers are.

At our facility, I carried on a tradition that was begun initially by the first Chief of Public Safety I worked for, Robert D. Jacoby. Bob, who was a "Fire Chief" himself of a large Long Island Fire Department felt it was important to have a working relationship with the people we were going to be depending on in an emergency. This of course, started with the fire department. Every year, right after the local department's elections, all of the Department "Chiefs" would be invited to our facility for an introductory meeting. This is where we could put faces to names. We would distribute updated site maps of the campus and also update the key rings containing specific keys for access. We would also set up facility walk-throughs for them as our main facility was a 20 story high-rise building. We also had twenty-seven outer buildings on a campus of approximately sixty acres.

This gave the new chiefs an opportunity to see the inner workings of the facility up close and personal. They were able to see where our engineering department was located, our grounds department, our flammable liquid storage area, pathology & labs etc.

It was a practice that I would maintain for the fifteen years that I was Corporate Director of Public Safety and Emergency Management. This worked extremely well both for us and the fire department.

They appreciated the access to the facility we provided to them and we appreciated their support and response when needed. This was also the time we would discuss possible mutual training exercises for the coming year. On many occasions, the fire department ran exercises on our campus utilizing one of our outer buildings and well as our tiered parking garage.

Sometimes we would join them in conducting the exercise, and other times we merely secured an area for them and observed. But it was all mutually beneficial.

Another way in which we assisted the fire department was to allow and assist them in conducting a communication site survey throughout the campus. This helped them immensely by going through each building, floor by floor to ascertain where there radios worked and where they didn't. They were also allowed to post signage for their department members as a visual cue for these radio transmission points. Again, the intent here was to make their response in time of emergency a little bit safer for them.

Another thing that we instituted at our facility, again under the direction of my former boss, mentor, colleague and friend Bob Jacoby, and continued on through my tenure, was to label not just our exterior buildings that warranted it, but also all of our interior research labs with the NFPA 704 placarding.
The NFPA 704 Standard is the System for the Identification of Hazards of Materials for Emergency Response.

This standard presents a simple, readily recognized, and easily understood system of markings (commonly referred to as the "NFPA hazard diamond") that provides an immediate general sense of the hazards of a material and the severity of these hazards as they relate to emergency response. A number rating system of 0-4 is provided to rate each of the four hazards and is placed on a placard and provide emergency responders with the information they need to determine the immediate actions to be taken in an emergency. Tables in the standard provide the criteria for the ratings and placard specifications such as letter size and arrangement of numbers and colors are provided in the standard. By doing this we provided valuable information to the firefighters who would be responding to our facility. A list of each location where the placarding was used was issued and updated annually to the Fire Department. Every bit of knowledge that we provided to those first responders made it that much safer for them when they were responding to our facility.

Another thing we had as an advantage was that several members of our Public Safety Department as well as other employees throughout the facility were also members of the local fire department. And many more belonged to other community fire departments. This created built in liaisons with our local fire department community. Identifying facility employees who are in the fire service or EMS is definitely something you should look into. Many facilities incorporate these employees into their Fire Evacuation and Emergency Response Program. On occasion they are even given a title such as "Fire Warden".

 As far as the Emergency Medical Service (EMS) responders, most community fire departments also had their own EMS crews. In addition to that, the local County Police Department also had an ambulance bureau.

Then there were the voluntary and private ambulance companies that also responded to our facility. And finally, we had our own Ambulance Department which always proved to be an invaluable resource to not only ourselves but to the community at large as well.

Great efforts were made through the years to make all of these responders feel comfortable, respected and part of our overall facility community. This included, linen and supply exchanges or refills, a small coffee room lounge set up in our Emergency Department where they could take a short break, complete their paperwork and have a cup of coffee. At various times EMS staff were also given tours of the facility so that they could familiarize themselves with the different areas and their assigned use throughout the facility. The EMS community was also always included anytime an Emergency Management exercise was planned at our facility as they were an integral part of any response, staged or real life incident.

On a final note, our facility during the years of my tenure was the hosting facility for the County Regional EMS Council which met monthly and represented all the various interests of EMS. This was another way in which we worked closely with our EMS responders.

At this point I'd like to discuss the "Law Enforcement" community, and what we were able to accomplish with the various agencies to make their potential response, safer for them as well. First and foremost, as mentioned earlier in this book, is the fact that you should know who the players are and be able to put a face to a name at the very least. You accomplish this by reaching out to your local sheriff's department or your local police precinct. Find out who the Sheriff or Commanding Officer is, introduce yourself and ask to set up a meeting.

Once you have your meeting you can find out exactly what resources these departments have that could benefit you during a disaster response. Conversely, let them know what you have that could assist them. Invite them to tour your facility and campus with some of their key officers or personnel. As with the Fire Department, the more information you can give them about your facility, the better able they will be to respond in times of need.

If they have a patrol car that covers your facility and area, ask the sheriff or commanding officer if you can meet the patrol officers assigned to these vehicles and introduce yourself. This could actually be done through your Security or Public Safety Director. Let them set it up but make sure that you're part of the whole process. Again, these will be some of the people you'll be relying on in a disaster response to your facility.

The county that my facility was located in on Long Island, New York was fairly large and had a big, diverse population (1.3 million people). As a result, we had a large multi-functional County Police Department. One of the divisions that we had a fair amount of contact with, was the "Bureau of Special Operations" (BSO). Another was "Emergency Services". These are the group of individuals who did extrication, rescue, and also include their "Special Weapons and Tactics" (SWAT) team. These were the big boys on the block, when they responded, you knew it was an emergency. Think of them as the "Cavalry" riding to the rescue like in an old time western movie. One of the areas we were able to provide some additional safety to them was through on-site training at our facility. Their specialized division personnel were able to conduct simulated elevator entry and rescue in our buildings which proved invaluable.

On several occasions we actually needed to call them for a response to remove people and patients stuck inside our elevators. They also had the opportunity to go inside our tower elevator shaft to see the inner workings and plan on how they would respond if needed. They were also able to exercise their rappelling skills both in the facility (elevator shaft) and outside of our buildings. Our relationship worked so well with them in fact, that at one time five different divisions of the Police Department were housed on our campus in different buildings.

To this day, their Emergency Ambulance Bureau and their Emergency Medical Service (EMS) Academy is still housed there. We also maintained very close relationship and good rapport with the "Office of Emergency Management" which at the time was run by the Nassau County Police Department. The Commanding Officer, Deputy Inspector John Carlsen, assisted us greatly in our mission. John, who has since retired from the Police Department, has become a well-respected and recognized expert instructor and lecturer in the field of Emergency Management, both nationally and internationally. In fact, he was instrumental in helping me transition from Healthcare to instructing in Texas A&M University's TEEX program. To this day, he remains a mentor, confidant and close friend of mine.

Not everyone will have the same opportunities or success that I and my predecessors had, working with your local law enforcement agency. But that doesn't mean you shouldn't be trying. It's all about building relationships with them and the rest of your community's first responders. Find out who the main players are and at the same time make sure they know who you are. Let them know you are a professional willing to work with them and someone who can be counted on during a disaster or emergency.

One of the occasions where Police, Fire and EMS assistance proved invaluable to us was in 1990 when "Avianca flight 52" crashed on Long Island. Avianca Flight 52 was a regularly scheduled flight from Bogotá to New York via Medellín, Colombia. On Thursday, January 25, 1990, a Boeing 707-321B, ran out of fuel on approach to John F. Kennedy International Airport, resulting in the aircraft crashing into the small village of Cove Neck, New York on the north shore of Long Island. Eight of the nine crew members and 65 of the 149 passengers on board were killed. Seventeen survivors were brought to our facility. In addition to this, all deceased victims were brought to the County Medical Examiner's Office which is located on our campus and attached to the main tower building. This proved to present some of the greatest challenges to us as once viewings had begun at the Medical Examiner's Office (M.E.), they became overwhelmed with the large influx of family members.

The Police at the request of the M.E. shutdown and secured the building. The entire process of I.D (Through photos only) was transferred physically to our facility in an area adjacent to our public auditorium. The auditorium itself was set up as a press and family center. This I.D. process continued over the next two days and strained the resources of our facility and that of the assisting agencies as well.

Police command posts of both Nassau County and New York City were set up in our Emergency Department parking lot. This was one of the times that we ran a separate EOC for the hospital alongside of both the NCPD and NYPD operations. This ended up being one of the best things we did, even though conventional wisdom calls for a "unified command".

It gave us much more flexibility in fulfilling our function while the Police concentrated on victim transfer and the actual crash scene itself.

All of the first responders, in addition to transporting both survivors and victims' to our facility also assisted in traffic and crowd control as the scene at our facility became chaotic with emergency vehicles, staff, responding family members and of course the press. It was an instance where all of the pieces came together and aided us in our ability to provide health care services during an emergency and disaster. The fact that the majority of first responders knew the physical layout of our Emergency Department and property proved invaluable and made their response more efficient and safer for them.

On another occasion, in 2005, the first time we had ever gone on to full facility lock-down, was when two armed bank robbers, being pursued by the police, came onto our campus and entered our facility. These individuals, after having robbed a bank two towns away, proceeded to lead the police on a high speed chase ending as they entered our campus. The individuals left their vehicle and fled on foot. With uniformed police from the County P.D. as well State Police swarming onto the property it was several minutes before we actually knew what was occurring. Once we knew however, we activated HICS and went on a full facility lock-down. The local High School across the street and the County Jail next door also went on full alert and lock-down as well. Access to the facility was limited to one entrance for visitors and staff. Patients entered through the Emergency Department. All other entrances were secured and security staff were posted along with Police personnel throughout the campus. The police, were able to locate and apprehend the first suspect, who was hiding under a vehicle in our tiered parking garage. This occurred within an hour of the lockdown.

The second suspect was also eventually found and apprehended inside one of our outer buildings.

It was later determined that the suspect had entered through the back of the building though a door which was supposed to be locked for security purposes. The door had been propped open by an employee who wanted the convenience of not having to walk to the front of the building from the main tower. This seemingly inconsequential act caused grief, anxiety and panic as it allowed the suspect to bypass security measures established for the safety of all. It also kept the facility on total lockdown for over three hours. The fact that Emergency Services, Bureau of Special Operations and Special Weapons and Tactics had performed several exercises over the years at our facility and campus gave them firsthand knowledge of the inner workings of it. This included the access tunnel system which runs throughout the entire campus. This in turn made it a little easier and safer for them to perform their dangerous mission.

We were extremely fortunate in the outcome. Both suspects were apprehended without gunfire or incident and no one at the facility was injured. As I said earlier, we'll be depending on these people and their response to our facility during a disaster. As far as it's within your power, make sure they know what they are responding too, where in the building they are going too, what they'll be exposed too and are safe as possible when doing so.

## Chapter 8

## Responding when "Disaster Strikes"
## Case history

In this chapter, we will take the time to look at several hospitals which were impacted severely and took direct hits to their institutions from natural disasters. We will look at their preparedness, their response and their resilience in the light of facing such a catastrophic event. We'll also examine their ability to provide services (if they were) before, during and after the event. Any finally, we'll see what the aftermath brought for these facilities, and look at their post disaster status.

We're going to start with one of the worst and costliest natural disasters ever experienced in America. I'm referring of course, to *Hurricane Katrina*, which hit the Gulf Coast in 2005.We are all aware of the devastation that hit New Orleans, Louisiana, but some of the worst damage actually occurred along the Gulf Coast. In particular, Alabama suffered severe property damage to their entire coastal area.

But for the purposes of illustration in this book, we will be focusing on what occurred at a particular healthcare facility in New Orleans. Specifically, we will be looking at "Charity Hospital" which was directly impacted and had to facilitate a full evacuation during the hurricane. The reason that I'm using Charity Hospital is that although many healthcare facilities were affected by Hurricane Katrina, only Charity Hospital failed to reopen its doors. First, a little background and history on Charity Hospital.

Charity Hospital was founded on May 10, 1736, by a grant from the French sailor and shipbuilder *Jean Louis*, who died in New Orleans the year before. The first Charity Hospital was located on the intersection of Chartres Street and Bienville Street in what is now the French Quarter. The hospital was founded 18 years after the city was founded by France in 1718. It is the second oldest continually operated public hospital in the United States. Only Bellevue Hospital in New York City is older, having been founded over a month earlier, on March 31, 1736.

In an ironic and strange twist of fate, Bellevue Hospital in New York City would also be struck directly by a hurricane (*Hurricane Sandy*) seven years later in 2012 and suffer severe damage as well as instituting a patient evacuation. But the outcomes for these two facilities would be completely different. More on that, later in this chapter.

As America entered into the 20th century, the city of New Orleans was rapidly expanding, and the demand for indigent medical services again exceeded the capacity of the existing Charity Hospital. The hospital, which had suffered through so many crises in the past, had also outgrown its own facilities capacity many times. So In 1939, what was to become the current, sixth and final incarnation of the hospital was built on Tulane Avenue. At the time of its construction, it was the second largest hospital in the United States with 2,680 beds.

"Before *Hurricane Katrina* struck in 2005, Charity Hospital was considered the pride of New Orleans. Charity Hospital was regarded as one of the most vital and successful facilities in New Orleans. "Charity was one of the best teaching hospitals in the country, where students from Tulane and LSU did their training," says Dr. James Moises, a former Charity emergency room physician.

He noted that it served 100,000 patients a year before the storm. When *Hurricane Katrina* hit Charity Hospital, there were about 200 patients and doctors who were still trapped in the facility under deplorable conditions. Eventually, it was fully evacuated of all patients and personnel.

Eighteen-wheel trucks had to be brought in and rolled up in front of the hospital to remove the final patients and personnel late on a Friday. The trucks went to New Orleans' international airport or to the state capital of Baton Rouge. The situation at Charity had become dire since *Hurricane Katrina* slammed into the Gulf Coast. The hospital had no power, no water and no food. Some patients were on ventilators being worked by hand pumps; the bodies of those who died were stored in stairwells, as the hospital's morgue had flooded.

 Although the damage was severe, Charity Hospital actually flooded only in the basement during Hurricane Katrina. And this flooding actually occurred after the storm hit and when the levees failed. The levees were mitigation measures constructed by New Orleans years earlier to protect the city from flooding as it lies below sea level in a bowl between the Gulf of Mexico and Lake Ponchartrain.

In an extraordinary act of dedication and volunteerism, a 200 member medical and military team brought in a 600-kilowatt generator, pumped out the water and prepared the hospital for service. The hospital was cleaned (to a condition better than before the storm) and was "medical ready" within several weeks, according to doctors and military personnel present at the cleanup, as well as Lt. Gen. Russel Honoré, the retired Army general who was commander of the joint task force on *Hurricane Katrina*. (This according to an article by Roberta Brandes Gratz, in "The Nation".)

But in spite of this, Charity Hospital, though it had sustained severe damage, was then cleaned to pre-hurricane conditions, and was determined sound by a national architectural firm, was never reopened.

For a while, there were some interim medical services being provided by Charity as overseen by LSU, but these did not last very long as other facilities took on a greater role in providing that care. Even the power which had been put back on to air out the facility was later turned off and the building secured barring the volunteers access to the building to continue their work.

Why did this occur to a facility that had served the community since the 1700's, was highly regarded and was considered an integral part of New Orleans Health Care system? Primarily it seems, Charity Hospital became a victim of money and politics. As a safety net hospital it served a large indigent and uninsured patient base. And while they were ready to re-use the facility and open it again for services, other forces were working against it behind the scenes. Although there were many conflicting stories and participants and many disputed claims, the general consensus of opinion seemed to be that a new facility in a different location would be better for the entire community. Hopefully, once the dust settles and a new facility is built, the residents of New Orleans, including the needy and destitute will still be receiving quality healthcare as they had in the past.

What is indisputable however, is that through almost 270 years of service to the community, a citywide fire, hurricanes, the Civil War, smallpox, yellow fever and AIDS, Charity had always been open to New Orleans' neediest patients.
The prospect of that rich history coming to an end, in a city already reeling, is disheartening to many in New Orleans.

Whatever the ultimate reason was for deciding not to reopen Charity Hospital, it was truly an end to a long and illustrious run. Therefore, there were no plans carried out to install mitigation measures. And as a result, even though Charity Hospital survived *Hurricane Katrina*, it seems it couldn't survive the inevitability of progress and politics both in the city of New Orleans and the state of Louisiana.

On a personal note, I had the opportunity to visit New Orleans in the summer of 2013, some eight years after the hurricane wreaked its havoc on the city. I toured some of the hardest hit areas. Whole tracks of land are still desolate. Many houses are abandoned or completely overgrown. And while many areas have been totally rebuilt, it's obvious, and it seems to be a reality that many of these areas are never coming back to where they were before *Hurricane Katrina*. And some are never coming back, period.

On to our next case study. In this one we will look at the 2011 tornado that hit in Joplin, Missouri and devastated not only St. John's Mercy Regional Medical Center, but the city of Joplin, Missouri as well. We'll see how the staff responded along with the community they served in getting medical services up and running to remain an asset to that community during this time of crisis. We will also look at what they did on an interim basis and how that evolved into the building of a new facility. You'll see that this would be a much different community approach than the one we just examined.

Again, a little bit of background on St. John's Mercy Regional Medical Center. The hospital was founded on October 24, 1896, by Appoline A. Blair. Appoline Agatha Alexander Blair (September 14, 1828 – Sep. 5, 1908) was a Missouri philanthropist, hospital founder, and wife of Senator (and Civil War general) Francis Preston Blair, Jr.

In 1878, after losing two children to illness, Blair gathered a group of 20 prominent women and organized the St. Louis Children's Hospital, for which she served as the first president of the Board of Managers. In addition to the St. Louis Children's Hospital, she is also sometimes credited with the creation of the St. John's Medical Center in Joplin, Missouri, in 1896. The facility was expanded in 1968 to include two connecting buildings of seven and nine floors respectively.

On May 22, 2011, the hospital took a direct hit from an EF-5 tornado (winds greater than 200 mph) and received devastating damage from the May 2011 tornado outbreak. A total of six people were killed inside the hospital, five patients and one visitor. Surviving patients were evacuated from the health facility and its grounds, which sustained major structural damage. One of the hospital's towers took a direct hit by the storm and was rotated four inches on its foundation. The hospital actually had video of the tornado hitting inside the facility. In addition its "Lifeflight" helicopter was thrown off the roof like a toy and totally destroyed. This all took less than a minute to occur (approximately 45 seconds).

Mere hours after the deadliest recorded tornado in U.S. history hit Joplin, the Missouri Disaster Medical Assistance Team (DMAT) started working to create a plan to help the survivors. The following Wednesday, DMAT deployed their 8,000 square foot field hospital to temporarily replace the destroyed hospital. Six days after the tornado hit, on May 29, 2011 St. John's medical staff gave medical treatment to their community in the BLU-MED field hospital.

The existing hospital was structurally unsafe and was eventually demolished. One week after the tornado, St. John's (now known as Mercy) announced they would rebuild. Mercy is in the process of building a hospital in Joplin at Interstate 44 and Hearnes Boulevard; it is scheduled to open in 2015, replacing the facility destroyed by the tornado. There will also be an auxiliary facility on the northeast side.

May 22, 2013 marked the second anniversary of the devastating EF5 tornado (winds greater than 200 mph) that flattened Joplin, MO, and left 161 people dead and injured more than 1,000 people. The town of about 50,500 honored its dead and commemorated the catastrophic event at a public ceremony that included a moment of silence at 5:41 p.m., marking the minute in 2011 when the tornado first struck.

It had touched down on the western edge of the town and began its devastating path of destruction. Joplin City planners used the event to highlight the progress that has made in rebuilding and to subtly shift the focus toward the future. Rebuilding and looking to the future are also the focus at the new Mercy Hospital Joplin, which is planning a March 2015 opening for the $450 million, 900,000-square-foot, 260-bed hospital that will replace the storm-gutted St. John's Regional Medical Center.

Despite the quick thinking and valiant efforts of St. John's staff two years ago, five patients and one visitor inside the hospital died from injuries and other storm-related factors when the tornado blew out windows, peeled off the roof, and shut down electrical power, back-up generators, and communications.

Immediately after the storm a team of engineers used the unique circumstances provided by the catastrophe to comb the gutted hospital and identify the weak links.

"Dr. D. Sean Smith, president of Mercy Clinic-Joplin (Mo.) and Kansas Division, said that, in his more than 30 years in emergency medicine as both a paramedic and physician, he knew the importance of hospitals participating in disaster drills, but he never was part of a drill where the hospital itself was the scene of a disaster." That is what occurred May 22, 2011, when Mercy St. John's Regional Medical Center in Joplin took a direct hit from a tornado that leveled most of the town. The 347-bed, nine-story hospital somehow remained standing, but the 200 mile-per-hour wind blasted through the windows destroying everything in its path. The tornado killed 161 people, including five patients and one visitor at the hospital. Sorensen credits the hospital's "hero nurses" who evacuated the patients from the building in darkness and driving rains for keeping the death toll from being higher. "The wind came in the west and blew out the east," Donn Sorensen Chief Operating Officer said. "I can't convey to you the level of destruction. "There were 183 patients inside the hospital that Sunday evening, including 24 in the emergency department and one that was in the middle of an orthopedic surgery. The tornado hit soon after a "Code Gray" (tornado alert) had been initiated, in which patients in exterior rooms were being moved into the interior, though Smith said it wouldn't have mattered even if they had time to complete that mission because "every room in the building had storm damage." In fact, the entire facility suffered damage from the tornado.

Once the tornado had passed, the toll it had taken on property and lives had become apparent. The injured residents of Joplin saw the hospital still standing and started heading toward it, for care and refuge, not knowing that its interior--which was filling up with leaking natural gas--was uninhabitable. "They looked like zombies coming in," Smith said. In spite of this devastation, staff who were uninjured culled together as many supplies as possible and set up a make shift care station.

This was set up outside in the parking lot. It wasn't long however, Smith said, that the medical staff on the scene were running short of antibiotics, painkillers and tetanus shots, as well as sutures and dressings.

By the following week, a temporary mobile hospital was established for 200 patients in the gymnasium of the local civic center, where Smith said they practiced "ditch medicine" and treated everything from minor scratches to a cardiac arrest. There was one bit of good fortune for Joplin in that a disaster drill was being conducted two hours away in Branson, Mo., where a temporary hospital had been assembled as part of the event. The facility was quickly disassembled and put back together in Joplin, complete with a 20-bed emergency department, catheterization laboratory and diagnostic-imaging equipment. "They were not the nicest digs I've ever been in, but we were damn glad to have it," Smith said, adding that the other bit of good fortune was that Mercy had just completed a system-wide electronic health-record rollout a few weeks earlier.

 This would allow the medical histories of residents who were patients of the Mercy system to be accessible. Smith said Mercy kept issuing paychecks to the hospital's employees and offered "turn-key" integration contracts to independent physicians who had practiced at the hospital. While this started as an "altruistic" initiative, Smith said it was also an excellent business plan because staff stayed in the area, unlike in New Orleans where "staff had dissipated to the four winds" in the aftermath of Hurricane Katrina and were unavailable when some facilities were reopened or replaced.
He added that it was a big boost to morale when, on Aug. 18, 2011, a new temporary "modular" facility opened that had air conditioning, running water and indoor toilets.  A two-story, temporary, concrete-and-steel 180-bed structure was opened in April of 2012.

This structure is planned to serve the community until a permanent replacement hospital is completed. Smith said a new 327-bed hospital (with room to expand to 424 beds) has been designed and is scheduled to open in the first quarter of 2015.

Among the lessons that were learned or reinforced, Smith said, were the need to have a secure "bunker" for communication equipment and medical supplies, that having mobile emergency medicine personnel treat injured patients at their location was more efficient than trying to transport them all to hospitals, and that it's vital to maintain good relations with other hospitals in the region--even those considered competitors (from an article by Andis Robeznieks, in Modern Healthcare). The new facility as a mitigation measure will be built to exceed the existing building code. A decision was made to go beyond the code requirements and not just meet them or their minimums. This included installation of windows throughout the facility that can withstand winds of up to 200 MPH.

And so, even though it was devastated by the tornado, St. John's Mercy, not only regrouped, but maintained services to the community that had been devastated as well. And with the help and support of that community is continuing to rebuild alongside that very same community to remain an integral part of its infrastructure. And as they move ahead together, it will continue to remain a valued asset by everyone in the city of Joplin.

Moving on to our next case history event, we'll look at two hospitals impacted by *Hurricane Sandy* in 2012. One on Long Island, New York and the other one in New York City.

Before I mention either of the hospitals or provide background on them I'd like to list some background data for

*Hurricane Sandy* itself. *Hurricane Sandy* (also unofficially known as "*Super Storm Sandy*") was the deadliest and most destructive hurricane of the 2012 Atlantic hurricane season. It ended up also being, the second-costliest hurricane in United States history. While it was still a Category 2 storm off the coast of the Northeastern United States, the storm became the largest Atlantic hurricane on record (as measured by diameter, with winds spanning 1,100 miles).

Estimates as of March 2014 assessed the total damage to have been over $68 billion (a total surpassed only by *Hurricane Katrina in 2005*). At least 286 people were killed along the path of the storm in seven different countries. In the United States, *Hurricane Sandy* affected 24 states, including the entire eastern seaboard from Florida to Maine and west across the Appalachian Mountains to Michigan and Wisconsin, with particularly severe damage in New Jersey and New York. Its storm surge hit New York City on October 29th, flooding streets, tunnels and subway lines, closing all area airports, waterways and cutting power in and around the city. This in effect, cut off Long Island from the mainland. Damage in the United States amounted to just over $65 billion dollars.

Now that we've set the scene for what actually caused all of the damage, let's look at two of the healthcare facilities in the New York area that were impacted by it. The two facilities we'll be reviewing are Long Beach Memorial Hospital in Long Beach, Long Island, New York, and Bellevue Hospital on the east side of Manhattan, New York City. Two entirely different facilities.

We'll begin with Long Beach Medical Center also know as Long Beach Memorial. Long Beach Medical Center (formerly *Long Beach Memorial Hospital*) is a teaching and community hospital originally licensed for 403-beds located in Long Beach,

New York. The hospital which was founded in 1922, also includes the Komanoff Center for Geriatric and Rehabilitative Medicine a 200-bed facility providing sub-acute and skilled nursing, founded in 1974, which is adjacent to the main hospital. The hospital lies right at sea level on the south shore coast of Long Island.

And while there have been flooding issues in the past both for the hospital and the city of Long Beach, nothing could have prepared them for what was to come. Even with all the mitigation measures that had been instituted over the years at the hospital, Long Beach Hospital was destroyed as a result of the ferocity of *Hurricane Sandy*.

As the storm approached, and its destructive power and intensity became clear, decisions were being made to protect its patients and staff. The Nassau County Executive ordered the evacuation of all healthcare facilities on Long Beach Island including Long Beach Medical Center and The Komanoff Center for Geriatric & Rehabilitation Medicine. Accordingly, the Medical Center began transferring hospital patients and residents from the Komanoff Center to other area hospitals and nursing homes. This was something that had been drilled many times over with hospital, city, county and state Emergency Preparedness professionals.

After the Nassau County Executive ordered the evacuation of all health care facilities in Long Beach on Oct. 28, Long Beach Medical Center transferred sixty (60) its patients to another large Medical Center in East Meadow.

The residents of The Komanoff Center were transported to a number of facilities on Long Island. Long Beach Medical Center initially established a command center at South Nassau Communities Hospital Training Center after the storm. (South Nassau Communities hospital would end up being a key player in the future of Long Beach Medical Center). The Medical Center's Emergency Department remained open until all of the inpatients were transferred and then it was closed along with the rest of the hospital. Unbeknownst to everyone at the time, was the fact that Long Beach Medical Center would never reopen in its present configuration again.

Within a week,  the city, in conjunction with the hospital and FEMA, had secured a federally staffed temporary hospital — DMAT (Disaster Medical Assistance Team) — located at a local baseball field, with a full service emergency department(the other DMAT team assigned to Nassau County was based at Nassau University Medical Center, my former employer). According to the city of Long Beach, DMAT saw more than 1,000 patients, including those who had sustained hurricane recovery-related injuries and illnesses. The Red Cross which was also on the scene, provided mental health counselors and sheltering needs for hurricane victims.

The hospital's basement was flooded in the storm, requiring officials to move the pharmacy to the third floor.  An inspection by the Nassau County Fire Marshal's office determined that the building's entire sprinkler system would have to be replaced because of concerns that salt water may have corroded the pipes. In addition, massive sections of their electrical and mechanical systems had to be replaced as well. The entire facility had to be cleaned and repainted. This would all take a lot of money. By some estimates, it was going to require at least $56 million in repairs.

Unfortunately, money was the one thing that Long Beach Memorial hospital did not have readily available, having run in the red for the past several years. The hospital which was licensed for 403 beds, was only operating at 162 beds.

This ultimately, would also play a big role in the hospitals inability to reopen. The hospital which was unable to make payments to its employees ended up being sued by a group of them for back pay and entitlements.

The New York State Health Department, recognizing the financial and well as care related issues that Long Beach was experiencing before *Hurricane Sandy* hit, denied the hospitals application to reopen unless major changes were made. The Health Department made the recommendation that Long Beach look into a merger possibility with another healthcare facility. Long Beach Medical Center, either unable or unwilling to initially follow this course of action eventually filed for bankruptcy over a year after the storm had hit.

Long Beach Medical Center which had been closed for almost a year, left hundreds of its employees without jobs when there was no hospital to run. Meanwhile the hospital, which used to employ 1,200 workers, was now down to 430. More than half of those now work in the Nursing Facility that was less severely damaged and was reopened. The one positive thing the bankruptcy did was pave the way for the potential merger with South Nassau Communities Hospital in Oceanside some seven miles away. In an agreement that was finally worked out between the state and the two hospitals involved; South Nassau Communities Hospital will acquire all the real estate and operating assets of Long Beach hospital and its affiliated nursing home the Komanoff Center at their appraised value.

South Nassau will continue to pursue FEMA funding to restore Long Beach's facilities, but whether the center will reopen as a hospital remains unclear. More likely, an emergency services facility will be opened in its place. The 200-bed Komanoff Nursing facility, which has about 100 residents, will continue to operate at its present location. And now, some four years later, there is still no healthcare facility open at the site of what was Long Beach Medical Center. And there may never be again, although community hope springs eternal.

The second hospital that we'll look at that was affected by *"Hurricane Sandy"* is Bellevue Hospital in New York City. We will look at the impact that *"Hurricane Sandy"* had on it, how it handled its response and how it recovered from that disaster. Once again, we will begin with a little history of the facility itself.

Bellevue Hospital Center (more often simply called Bellevue) was founded on March 31, 1736 and is the oldest public hospital in the United States. It has a reputation as being one of the best hospitals in the city. It handles nearly 670,000 non-ER outpatient clinic visits, over 99,000 emergency visits and some 26,000 inpatients each year. Like Charity Hospital in New Orleans more than 80 percent of Bellevue's patients come from the city's medically neediest underserved populations. Although it is a safety net hospital, it is also a hospital of *National and World* firsts.

In 1799, it opened the first maternity ward in the United States. By 1873, the nation's first nursing school based on Florence Nightingale's principles opened at Bellevue, followed by the nation's first children's clinic in 1874. The nation's first emergency pavilion opened in 1876; a pavilion for the "insane" (an approach considered revolutionary at the time) was erected within hospital grounds in 1879.

Bellevue initiated a residency training program in 1883; it is still the model for surgical training worldwide. The Carnegie Laboratory, the nation's first pathology and bacteriology laboratory, was founded there a year later, followed by the nation's first men's nursing school in 1888. By 1892, Bellevue established a dedicated unit for alcoholics. Bellevue opened the nation's first ambulatory cardiac clinic in 1911. It was also the site of the establishment of the first ambulance service. These are just some of the many National firsts.

Bellevue was also the site of several world firsts. By 1867, Bellevue physicians were instrumental in developing New York City's sanitary code, the first in the world. In 1941 Bellevue became the site of the world's first hospital catastrophe unit. One year later, the world's first cardiopulmonary laboratory was established at Bellevue. Again, this is just a short listing of many, many firsts. I just wanted to give you some idea of the wide range of Bellevue's history and services.

After Hurricane Irene hit in 2011, the hospital, located near the East River in Manhattan, took steps to protect its operations against future storms. The hospital instituted many different mitigation measures. One of the things Bellevue did was to build a special encasement for its fuel pumps, equipped with airtight doors. The hospital's bulk oxygen and nitrous tanks were similarly protected. In addition, in a move performed earlier, the emergency generators had been relocated on a much higher level floor (13) of the hospital.

These protective measures were based on flood mapping data available at the time from the Federal Government. When Hurricane Sandy struck Oct. 29, 2012, it became clear that the historical data provided, was no longer accurate or sufficient.

The storm surge from Sandy however, overwhelmed the hospital's defenses, filling the 182,000 square foot basement with water that ranged in depth from four to eighteen feet. The basement housed over 200 pieces of critical equipment, including computers that were the mainstay of the hospital's IT department, bulk oxygen and nitrous oxide tanks, electrical switches, heating, air conditioning fuel pumps, and water pumps. Some of this equipment was salvaged but most was destroyed.

The pits for the hospital's thirty-two elevators were also in the basement. As a precaution, the elevator cabs had been parked on the first floor before the storm. After Sandy, all of the elevator cabs were able to be cleaned and reused; the cables and some electrical equipment were replaced.

When the main power went out, the hospital switched to its emergency generators, which are kept on the 13th floor precisely so they don't get flooded.  But as the water continued to rise, the fuel pumps electrical transfer switches in the basement shorted out and stopped working, the New York Police Department brought in a fuel tanker. Hospital staff, volunteers and eventually hundreds of troops from the *National Guard* spent the next thirteen hours carrying five-gallon containers of fuel up thirteen flights of stairs so the generators could continue to operate. The next day, the water pressure started to deteriorate and a decision was made to evacuate the entire hospital. Originally, there were 725 patients -- including prisoners and psychiatric patients -- in the hospital when the hurricane hit and flooding began. About 500 patients were eventually evacuated, and approximately 225 were discharged. The most critical patients and all but a few infants had been transferred to other hospitals before that Wednesday when it finally shut down.

Bellevue Hospital Center, New York City's flagship public hospital and the premier trauma center in Manhattan  worked into the night to so that the remaining  patients left in its darkened building were to be evacuated and transferred to other facility's on Thursday.

Health and Hospitals Corporation, which runs Bellevue, and all of the other Public hospitals in New York City, described the conditions as third-world like. There was no hot water, no lab or radiology services and pails of water had to be hauled up the stairs to use for flushing toilets. After pumping out 17 million gallons of water from the basement, the water was still two and a half feet deep in the cavernous basement of the facility.

The Health Department authorized the city hospitals to activate their "surge-capacity plans" that allows hospitals to accept patients beyond their normal capacity in a disaster, and if necessary, converting nonclinical space like conference rooms and auditoriums into dorm-style wards. Appeals were made for other hospitals to take Bellevue patients went out at midday and were immediately answered. One of the great advantages of New York City, and the greater metropolitan area, is having so many quality healthcare facilities to assist each other during times of disaster. And this time was no exception.

In January 2013, FEMA issued updated Advisory Base Flood Elevation (ABFE) data, reflecting new flood zones as a result of *Hurricane Sandy*. Bellevue is making plans to use the new data provided by FEMA to protect itself and its patients from future storms. Bellevue began planning for future storms soon after Sandy, using the latest FEMA data.

Bellevue then instituted the following additional mitigation measures. These were in addition to the ones previous put in place after *Hurricane Irene* hit in 2012.

- Many of the electrical and mechanical systems that flooded were replaced on higher floors and platforms were built to elevate those systems above the 500-year flood level.
- Selected elevator pits are being moved to the ground floor so they can function during an emergency.

- The emergency power distribution system is being expanded to bring generator power to key areas of the hospital, including sections that house CT scanners and MRI machines, pharmaceutical and chemotherapy facilities and research laboratories.
- Options for water pumps include moving them to a higher floor or bringing in additional pumps at street level that can be used as a backup system.
- Engineering experts are looking for ways to improve protection for the fuel pumps and medical gas tanks.
- The hospital is adding connections for mobile boilers that can be brought in to distribute steam to provide heat and hot water if necessary.

This project serves as a model for protecting against future disasters. It shows just what can be accomplished in the face of adversity. Mitigation involves educating the community, protecting the environment, providing risk management and taking steps to rebuild stronger and safer. It allows for the facility to remain an asset and part of the critical infrastructure of the community it serves. In this case, a major healthcare facility was inundated by a disaster, continued to provide healthcare for as long as possible, and when unable to continue its mission, evacuated its patients and moved its staff to safer locations.

It then showed its resilience and determination by fully reopening its doors for service, the first week in February, 2013, just a little over two months after it was impacted by *Hurricane Sandy*. And while improvements and mitigation efforts continue, Bellevue continues its long history of service to the people of the City of New York.

For the final case history in this chapter, we'll be focusing on Moore Medical Center located in Moore Oklahoma. On

Monday, May 20, 2013, Moore Medical Center received a direct hit from an EF-4 tornado. The facility was totally devastated and pretty much rendered useless. The amazing and maybe even miraculous thing about this disaster is that none of the thirty patients or accompanying staff were killed or injured. And of the 250-300 community residents that sought shelter at the hospital only minor injuries were reported among them. At the end of it all, the building was gone but everyone was still there.

Moore Medical Center was a forty-six bed acute care hospital serving a community (Moore City, Oklahoma) of over 55,000 residents. They averaged approximately 35,000 Emergency rooms visits annually. After the tornado struck, staff immediately regrouped and continued with patient care. Their first priority was to make preparations to transfer and move the approximately thirty patients from within their hospital to other healthcare facilities.

One of the first obstacles they encountered, was that the main entrance of the facility was pretty much impassable due to all of the debris caused by the tornado. This did not stop them, but simply slowed them down as they adapted and found ways to get everyone out safely through a rear entrance.

Next on their agenda was to provide as much care as possible to the residents that had made their way to the facility with various injuries. A total of fifty-one residents lost their lives due to the tornado.

Within weeks of the disaster, a company that supplies and installs temporary medical facilities was in touch with them to assist their needs. This same company had also provided facilities in Joplin Missouri and New Orleans, Louisiana to those healthcare facilities when they faced their disasters.

Eventually, five units were erected and interconnected to serve the community of Moore, Oklahoma.

Plans are currently in the works to rebuild a new, five story twenty-eight million dollar medical facility. Opening is set for mid-2016. Initially, after much feedback from both staff and residents, the facility will only offer extensive out-patient services. If needed in the future, in-patient services could be added as part of their phase two plan.

One of the most important mitigation efforts on their part that was instituted was to fabricate a tornado shelter directly inside the facility. In the event of a future occurrence, this will no doubt assist in protecting lives.

As you can see by the five case history's I've presented, each facility had a different and unique ending after their encounter with a disaster. Unfortunately, they would not all be happy endings.  But then, this is not a "Fairy Tale" but rather, real life, and as such, we must deal with the harsh realities that life brings. Things that are sometimes beyond our control. Two hospitals are rebuilding a new facility, one hospital reopened within months, one facility is in the process of a merger and one facility never reopened.

And so, even though facilities institute mitigation measures, conduct preparedness drills and exercises, write their policies and procedures, and train their staff, it still may not be enough when disaster strikes. Quite honestly, all we can do is continue to prepare, and respond accordingly when the inevitable occurs. A good adage for Emergency Managers to always keep in mind is to "Prepare for the worst, but hope and pray for the best".

## Chapter 9

## Are you viable, can you continue to function?

If or when your facility is ever impacted directly by a disaster, this chapter's title will be one of the very first questions that will have to be both asked and answered. Lives may very well depend on the answer we give or get. We need to, as quickly as possible, assess the damage (physical building, grounds, staff, and supplies and equipment) to determine the facility's ability to continue in its mission. That's where a well-run and well-staffed Emergency Operations Center (EOC) will prove invaluable to you. By being able to receive accurate and descriptive reports from your staff, your effectiveness becomes enhanced and your value to your facility and staff is magnified. Remember, your facility is a key critical asset to your community, and whenever possible should continue in its mission. But never, at the risk of the safety of your staff and patient population. This must *ALWAYS* be your first concern. Let me reiterate something I've stated several times earlier in this book, if you protect your staff, then they will be able to take care of others.

Alright then, you're facility has taken a hit; from flooding, a hurricane, tornado, fire or explosion to name just a few. You've started to receive some initial damage reports. Start by assessing your viability immediately along with your engineering staff and public safety personnel. Not only do you need to know how the exterior of the building and grounds were affected, but the interior of all buildings as well.

Some of the questions that you will need to start asking immediately are; what is the status of our infrastructure, do we have power and HVAC, do we have water, do we still have means of entry and egress to the facility? Are the roads passable in and around the facility? Have any of your staff, visitors or patients been injured, or heaven forbid, killed? When assessing any of the damage reports that come in, another thing to consider is how much of the facility was impacted. Was it the entire building(s) and campus or was the damage relegated to certain areas and buildings? If the entire building or facility is affected and severely damaged like the facility hit by the tornado in Joplin, Missouri, initially there's not much you can do, other than to secure the site and make sure that your staff and patients are out of harm's way. The time for repair, rebuilding and new construction will come later.

The task at hand, right now for you, is to evaluate the stability of the facility and determine its ability to continue in its mission, in a safe setting for staff and patients alike. What about your outlying buildings, did they suffer damage also? Can any of them be utilized? And as stated before, this needs to be done as quickly as possible.

Once you've established (if possible) that the entire facility is not damaged, determine from your status reports which units and areas are still capable of functioning. Then you can, along with nursing and the medical staff, ascertain how many additional patients you can shift from the areas affected to these safer units. The total number of patients you have will have been impacted somewhat by whether this was a "sudden event" or one in which you had some warning and were able to discharge or transfer patients.

Regardless, make sure that staff from the affected areas, as well as patient's medications and charts follow along with those patients to the units that are still functioning. This will provide both continuity of care for the patients and specific nursing expertise pertaining to the new patient's medical or surgical needs. It will also provide patients with some comfort during this chaotic time.

Mix patients as appropriate and with the authorization of the Medical Director or other appropriate title position. To accomplish this task, you'll actually have to perform several small scale unit evacuations throughout your facility. Utilize the same principles and procedures as if you were going to do this for the entire facility. Now is not the time to reinvent the wheel. Go with what's tried and true, it will save time and be much more efficient.

This means gearing up staff so that you have adequate and appropriate people to accomplish the task at hand. Make sure to emphasize with them just how important it is to accomplish this as quickly and safely as possible. There is always risk involved in evacuating hospital patients, whether they are elderly, young or non-ambulatory. Also, there is always the potential for fatalities (for any number of reasons) during evacuations. Don't let this delay your ability to provide a safer environment for the greatest number of patients possible. Remember, you may be doing this several times and in various areas of your facility. And time is always working against you. Act quickly where possible, but always mindful of the fact that safety for all is paramount.

In addition to your own facility staff, you will likely be able to enlist the aid of the first responders who will be arriving at your facility. Embrace both their response and willingness to help.

Again, this is where prior drills, exercises and joint training can reap their greatest value. If staff are able to assimilate quickly with each other, the entire evacuation process will be easier and proceed smoother. There will always be bumps in the road, but don't worry about them, instead, minimize them where possible and find ways to work through them.

One thing for you to remember, is not to get too hung up on who's in charge or who is directing staff. Work with the first responders and their supervising staff. Let them know what you need, what you have and how they can help. They'll work out the specifics and particulars. This prevents egos from being ruffled and avoids turf wars, especially at a time when cooperation is critical.

After you've moved your patients from the affected areas to secure ones, it's time to stop for a second and take stock of the situation. Has the full extent of the facility damage been evaluated? At this time, do you really know which areas are safe, reliable and usable? If you don't, you need to find this out *NOW*. This answer and information will help guide you, as well as your administrative staff in making any decisions moving forward.

Depending upon the type of disaster, and how much of the surrounding community was affected, will all be contributing factors to the amount of aid and how quickly it arrives from the community to your facility. Again, this is where your relationships with other healthcare facilities and vendors, especially ones from outside the affected area become critical.

Something to think about now and not at the time of a disaster is what is your facility's record and policy for paying vendors on time is. What happens if they borrow equipment or medications from other facilities?

Are the bills paid on time? Are the goods returned or replaced in a timely fashion? If they are, you should be in good standing. If not, you might find yourself near the bottom of the list, especially if other area healthcare facilities have been affected as well. Do whatever you can to prevent this from happening. Conduct some pre-planning in this regard and find out from both your purchasing director as well as your finance people what your facility status is, and what if anything they can do to improve it, if it needs it. If you have the opportunity, explain the negative impact (including additional or higher costs) it could have on the facility in a time of disaster. This is probably something they've never even thought about.

Having begun the process of moving your patients and assessing your facility damage, you should also have started the process of opening and utilizing your *"alternate care sites"* that have been pre-identified and surveyed for appropriate use. Certainly, if possible, you'll want to utilize any areas identified within your facility or confines of its property first. The simple reason for this is logistics. The shorter the distance of transferring the patients the better it is for everyone involved in the process. Utilize all areas you can as long as they provide a safe environment. Something else to consider is to think "outside the box" when it comes to opening your alternate care sites. Whatever you originally had them designated for may not be your first priority now. For what other purpose could they be better utilized?

 For instance, you may not need an area for maternity or pediatrics, but that area would work great for housing staff or even staging supplies that come in. These are the types of things and decisions you may be required to make on the fly as you respond to the disaster. The fact that you and your facility will be able to do that, will be one of the things that help determine how effective all of your planning and training has been. This will certainly be a "Reviewable event" for you when everything is over. Something else to keep in mind, while everything is swirling around you, is to remember to work with other staff and delegate what you can. In the chaos and intensity of the event, don't let yourself become overwhelmed. No one is truly effective after working under pressure for sixteen, eighteen or twenty hours. Please remember, that no one is indispensable, not even you. Your value to your facility is in your knowledge, expertise and professionalism.

The ability to function at a high level while "under fire" if you will, seems to be a common trait shared among many Emergency Managers. Maybe it's the nature of the profession, or maybe we're just built that way, regardless, it's something that serves us well during these times of need and crisis.

The next step for you, once you've activated your on-site alternate care sites, is to see if you need to utilize any of your off property sites as well. As you know this requires a lot more in the way of logistics (staffing, transportation, medications, supplies, hygiene etc.). If you have the need, don't hesitate in your decision if it might potentially save lives. Again, work through any issues you encounter. You'll want to also look at what kind of space you have and what you'll be using it for off-site. In addition, try and determine, if possible, how long you might need to use these off property sites.

The issue of staffing and resupply will be something you're going have to deal with, especially, the longer these sites stay open. Once again, you may want to think of alternate uses for them rather than what you may have originally designated them for. Remember that during a disaster, crisis or emergency, you'll constantly be evaluating your status and that of your facility as well.

A good Emergency Manager realizes the importance of remaining flexible in both thought and actions. You can't allow yourself to get bogged down by trying to follow a policy or procedure so much, that you lose sight of your end game. This is what you've trained for. You're a professional, use your knowledge, experience and even your gut feeling to your advantage to get you through.

During this time of assessment and evaluation of your facility and patient's needs and safety, please don't lose track of or forget about your biggest facility asset; your staff. I know I've spoken about this in the book already, but it is important and bears a reminder.

None of this will work without your staff's buy-in and involvement. They are our family, treat them as such. You're now at the stage where your facility's viability has been assessed and evaluated.  All of your patient's within the facility have been moved so they can be cared for in safer areas or have been transferred to other buildings and/or facilities. But what about receiving new patients? What is the status of your Emergency Department? Is it still viable? Do you have the specialized staff, supplies and equipment to adequately receive, assess, triage, treat and admit new patients? Or, do you need to consider treating and transferring them to other healthcare facilities not as affected by the disaster as yours. And in the extreme case, if your Emergency Department is not viable, do you need to consider whether or not to go on "diversion".

 While diversion is not something facilities do or take lightly (as it can cost them tens or even hundreds of thousands of dollars in lost revenue) it may be their best and only option. Two of the factors that have to be considered are staff and patient safety.

By trying to admit new patients, will you overtax an already stressed staff contingent, and as a result, will this subsequently put your existing patients at an even greater risk then they are already facing?

Either way, you'll now have to start focusing on the issue of resupply for your facility. This in the event that you are able to continue to remain open, even if it is only in a diminished capacity. Begin by conducting review and update with the department heads who are responsible for some of your essential services. Things like fuel levels for generators, food stocks and bottled water (if needed), pharmaceuticals, medical supplies, linens (laundry) are some of the items to name just a few. Remember, if you are going to be housing staff, and in some cases their family members or even people from the community your supplies will be used at a greater amount and faster rate. You will need to consider what your resupply availability and capacity is. This includes, available of product, ability to transport, and turnaround time for that resupply. Where are these products, supplies and equipment coming from? Much of this will be predetermined by what kind of disaster occurred and just how extensive the damage was to you and the surrounding community was.

As I mentioned earlier in the book, resupply routes for some of the affected hospitals on Long Island, New York were non-existent after *Hurricane Sandy*. Initially, roads, bridges, tunnels, airports and even waterways were closed or impassable. This negated the ability of suppliers from outside the geographical area from getting in to assist causing additional problems. Hopefully, you'll never have to deal with an issue like that.

There is something else I would like to take a moment to stress again, and that is the importance of understanding, and complying with "The Joint Commission's" 96 hour standard (If you use TJC). Even if your facility utilizes another accrediting agency, I truly feel this particular standard is important enough that you should familiarize yourself with it and use it as a planning tool. You'll be glad you did. Without going through it again here, just remember how important good planning is to you and your facility. It will pay off big dividends on the back end if ever needed.

There is another aspect of viability I would like to address in this chapter. What I'm referring to, is once everything has subsided and stabilized somewhat, how do start to affect repairs while maintaining a patient population. How do you tend to your patients, and where do your patients go during these repairs? Your facility may choose to start in the areas that are already vacant and closed. That obviously would be ideal and the best choice for you. But what if that is not feasible. What if repairs are also needed in the areas containing patients? What do you do then? How do you maintain patient safety throughout this process?

I would like to make a few suggestions to help aide you in this task. Hopefully, they might make the entire situation a little easier to deal with. First of all, is there any appropriate space within your facility to temporarily use as "swing space" for your patients? What about any of your on-site "alternate care sites"? Specifically, the ones located within your facility. Are they still being used, or can you reallocate them for this use? Remember, we're looking at a temporary move and transition. Hopefully, not more than a few days or a week. And it may not even involve all of your patients from an affected area or unit. Perhaps you may even be able to open a room or two and use them to shuttle patients around the repair schedule.

This would be an ideal situation as they would remain on their own unit with their own nurses and medical staff. Don't forget about the importance of patient "peace of mind" in the healing process. Have staff reassure them whenever possible that they are safe and will be taken care of at your facility.

Next on your list would be for you to look at your alternate care sites located on your campus but not in your main building. And then finally, what about the feasibility of using your off-site alternate care sites for transitional swing space? You should be thinking ahead on this issue before you consider closing any of the sites you may have already opened. Why would you want to go through the process twice if you don't have too? Even spaces like auditorium's, conference rooms or even gyms (if you have them) could be utilized for some ambulatory patients for a period of several hours while repairs are made. Obviously, the patient's medical condition would dictate the feasibility of this move.

These are the types of things you need to look at and consider before the disaster, during the disaster and then finally, after the disaster. In some cases, the after disaster evaluation may very well be the most important of the three.

For it's at that time, when you have the chance to evaluate how well (or not) your plans and procedures worked for you and your facility. This then, sets you up on the other end to adapt, and, if necessary, improve those plans moving forward. Our entire process is one of constant evolution. Nothing ever stays the same and no two responses are ever quite alike.

All we can ever really hope to do is to be part of and in some cases, drive that evolution process. In this way, we can be assured that we will be preparing our facility, for what is to come, to the best of our ability.

In a simplification, you can almost boil the entire decision process on whether or not you're viable and able to continue, down to three things;

Are your staff safe?

Are your patients safe?

Is the facility safe?

So, if you answer no to any one of these questions, you need to correct it, and if you can't, then you need to take appropriate action as quickly as possible.

In closing, I'd like to state that you may face many different issues and variables regarding your facility's decision on whether to remain open in the face of disaster. It is not an easy decision to make. It was never meant to be. No chapter or even book could adequately cover every single variable. In fact, in many ways, it may have just scratched the surface. Rather, the intent of this chapter was to get you thinking about the process, and some of the issues you will face in deciding what to do. It was also meant to be a prompter of sorts, so that after you've started thinking, you'll start asking questions. And then once those questions are asked, answers and solutions will follow. Remember the process is as much about teamwork as anything else. Finally, the answers you find will then help you to plan and prepare for any future inevitability that might arise. The rest as they say, is now up to you.

So, now that you've come along with me on this journey, let's make the final turn to bring it all together. As we move on to the final chapter of this book, I'd like to commend you all once again.

This is a difficult, but very rewarding profession to be part of. It's not for the faint of heart. Always take pride in what you do. The fact, that this is the profession that you have chosen to go into, speaks volumes about your character and who you are as a person.

# Chapter 10

## What did we learn, and what can we do better?

The quick and easy answer to the first part of this question is; a lot, and to the second part of the question, probably everything. Without a doubt, there is always room for improvement and ways of doing things better. The more complex answer however, is that devil is in the details. There is so much more to it than just a simple yes or no. That's what we'll look at and discuss here in this final chapter.

Let's begin by looking at the first part of this question. When we ask "what did we learn" we're actually looking for something tangible that we can measure or quantify as a result of our disaster response. Things like moving or evacuating patients is a good example. Response times of personnel are another. Were we able to complete these tasks in a timely fashion? Or did they require a much longer period of time than anticipated. If we transported patients, did we have adequate equipment and vehicles to do so? What about staffing levels to perform this task. Did we need more or less staff on hand to perform this task, and was that staff appropriately trained for the task assigned? What about our communications abilities? That is the one consistent area that people almost always feel improvements can be made. These are the types of things that are measurable. The answers to these questions then help us to adapt, evolve and become more efficient in the future. Much of what we do is based on historical data and previous responses.

It is only after an incident and then through our critiques, evaluations and after action reports that we truly learn from our response to a disaster, exercise or drill. Remember to always include both positive and negative observations when evaluating your response. If a portion of your program's response fell short of exception's or failed, find out why. Was it application, expectations or forces beyond your control that accounted for the breakdown or failure? Do your best to discover the root cause? Then after careful review, we can implement the appropriate changes to try and improve both our process and goals the next time.

Unfortunately, there will always to be a next time. But, then again, that's the business that we're in. It's always good to keep in mind, that every emergency or disaster we encounter becomes an opportunity to learn and to improve upon our future responses and our ability to provide a safe environment.

So let's continue to look at the evacuation example. Some of the things you'll want to evaluate are just how many patients you had at the time of the disaster and how many of them you actually had to move. This pertains to moving them both internally within the facility, and to moving them out of the main building to an on-site or off-site alternate care site. It will also cover evacuations and transfers to other healthcare facilities. Did you have a projected timeframe to accomplish this task? Were you able to complete it within that timeframe? If not, what were the reasons or obstacles you encountered that prevented you from accomplishing it? Were any of them avoidable or were any a direct result of failures within your plan or system? Was your plan adequate to deal with the crisis at hand or did it fail the facility at any point?

Regarding evacuation, these will be the type of questions that you and your review group will need to ask and answer yourselves. You literally have to pick apart your facility's response to the disaster step by step. This needs to be accomplished with an objective mind and without showing preference or favoritism of any kind. Don't ever gloss things over to try and make yourself or facility look good. You commit a disservice to yourself, profession and facility if you do.

Keeping that in mind, enlist the aid of your Emergency Management Committee or the group that helped you put together the facility "Hazard Vulnerability Analysis" to do this. Then ask yourself, did you accomplish what you wanted to do. After you answer that, then complete your review to see if you met your all of your goals and objectives.

This then, becomes your learning curve for moving ahead and improving upon the process. This process will need to be a multi-phased project. First, you'll want to review everything with the appropriate staff within your own facility. A nice touch here is, once you've completed your review and "After Action Report", put together a short power point presentation. Then use it as a review tool with administration, department heads and even individual departments. This way, everyone remains part of the entire process from planning, through implementation and response, right on to recovery. Remember, you always want to maintain staff participation and enthusiasm for your program whenever possible. Their active participation in the process gives them some ownership of the program as well. Also don't forget that, "The Joint Commission" (TJC) standards require that Administration be actively involved in the entire process of Emergency Management and not just briefed about it. By doing this, it helps you to be compliant with that standard.

After you've completed this phase, the next one you want to move on to is your first responders. Be sure to contact the Fire Department, Police Department, EMS and any other affiliated first responders and request a joint incident review with them. Make your report available to them and request feedback from them as well. This makes good sense and is an extension and continuation of the networking process.

It will also provide continuity for you, your facility and these agencies. A quick word of caution here, don't just present your findings to them. Solicit their observations and seek out their recommendations as well. Again, this gives ownership and lets them know that you value not only their physical response, but their intellectual one as well. Be professional at all times, just don't profess to them.

Conduct the same process for any outside agencies and/or vendors that assisted you during your response. Again, reach out to them for their input. The Department of Public Works, American Red Cross, and Salvation Army are all groups that have a stake in emergency response. If they were part of the response, see if they had any difficulties in their response to your facility. Look for feedback, and don't forget let them know how much you and your facility appreciated their assistance. Then let them know what you encountered, and together work on improving the response process in the event of a future disaster or emergency. You might even want to use your power point presentation as a learning tool for this purpose with them. Or, better yet, create a second power point presentation aimed specifically at first responders and outside agencies. Believe me, it will well be worth your time and effort. Once you complete this process for the first task (such as an evacuation) you then need to repeat the entire process for all of your other tasks, needs and objectives.

Supplies, equipment, food, power, water, staffing, housing, community support, network support to list just a few of the examples of things that were dealt with during the disaster response. Did you have pre-set goals or expectations prior to the disaster response? If you did, were you able to meet them, and were you able to measure or gauge the effectiveness of what you did?

This will then give you some new benchmarks to work from and measure against for the next time. The information and data that you gleam from this process will then go into the *"After Action Report"* (AAR) that you and your group will prepare and present. It will also be used to update and modify, if necessary, your *"Hazard Vulnerability Analysis"* (HVA).

Remember, the HVA, at a minimum, should be reviewed annually and after every disaster response or major exercise. This way, you will be working off of the most current and up to date information. Something to bear in mind, as was stated earlier, is that no Emergency Management Plan can be effective if it remains "static". It needs to be a living, breathing, evolving document. It is intended to help guide you in your response actions, not be the "end all" to them. As I stated earlier, don't ever become so focused on your plan that you lose site of the bigger picture. If, after all is said and done, you find that there were shortcomings regarding your response, don't become disheartened in any way.

Instead, use it as an opportunity to improve and as a learning tool. This is something I did on more than one occasion during my career. As a matter of fact, once I was aware of an identified issue, I would take it to the next step. Never be afraid to try to turn a failure into a success. This is all part of the learning process.

Again, this is something I learned from others who went before me. I would highly advise you to do the same. What I'm speaking of here, is to conduct an upcoming drill based on what you found lacking in your response. Then use it as the basis or focal point for the entire drill. Set an objective, then see if you are able to achieve it. If it's unachievable, analyze the reasons why the goals were not met. In that way it truly becomes a full circle learning experience.

This also shows both follow through and follow-up on your part and that of the facility. By doing this, you illustrate that you and your facility are committed to not only Emergency Management, but to quality improvement as well. I know this is being repetitious, but make sure you document this entire process completely.

This is the type of thing that "The Joint Commission" (TJC) and all of the other accrediting agencies like and want to see. It shows them that not only are you complying with their standards, but rather, that you actually understand the intent, application and purpose behind them.

Another suggestion I'd like to make, would be for you to use what you've learned as a bragging point if you will, with the accreditation surveyors. Explain the shortcoming(s) you identified during a response or exercise. Then speak about how you revisited it by conducting a follow-up drill and monitored and gauged the outcome.

Do the same thing to illustrate the things that worked better than expected as well. Many times, surveyors when shown these types of things will ask for a copy or permission to use as an illustration point in other surveys. By doing this, you actually end up teaching and helping people in the field that you don't even know.

That's kind of a bonus for us. Then finally, illustrate how those results were integrated into your quality improvement process, and used to update and improve your procedures, plans, process and program.

After reviewing your response, as well as your facility's resilience during the disaster, along with its recovery, you'll have a better picture as to your overall readiness. Then you'll be able to move forward and focus on what can be done better in the future, to protect your facility and staff and make it a safer place to work. But first, ask yourself, what was the overall impact not only to your physical facility and building but to your staff as well? Remember, we're not talking about physical injuries here. Your staff are part of your community as well. What were "THEIR" losses? Property, homes, family members (injured or deceased). They are as much part of your recovery as repairs or revenue. Maybe even more so.

Once you've done this, you'll want to enlist as much support as possible, from as many levels in the facility as possible, to accomplish this task. Try and look at it through many pairs of eyes, not just yours. You have a lot of knowledge, expertise and experience working at your facility. Now is the time to make good use of it. This is where the personal relationships you forged within your facility will reap the biggest benefit. With this information, you then need to revisit your Hazard Vulnerability Analysis, your Emergency Operations Plan and finally, your Comprehensive Emergency Management Program to make potential changes. There is no secret formula for this process. It just takes time, dedication and hard work to accomplish it. In this final chapter, I've attempted to illustrate some of the ways, and show you some of the things you can put into practice to foster a good solid learning process for your Facility Emergency Management Program, especially after a response.

I have used many of these myself and found them to be beneficial. And the ones that did not yield the results I expected were either modified or replaced with others. Much of this is trial and error. Don't be afraid to experiment, find what works best for you. Make sure that your facility recognizes and understands the importance of preparedness and the potential consequences for not making it a priority. One of the ways you can accomplish this task is by being a "good salesperson". Both to the people above you and for the ones you're responsible for. You, as the Emergency Manager, must know and understand the appropriate regulations and codes. Stress staff training, (remember Rick Rescorla) but do your best to make it interesting, informative and as enjoyable as possible. Keep in mind that you'll be dealing with adult learners who require a deft touch. Then, run as many exercises or drills as possible. By all means, try to remain positive and forward thinking. Continue to explore ways to keep your facility ahead of the curve.

Try your best not to be a reactionary manager. This book will provide a good base for you to start from. Feel free to tailor any of the presented material to your own needs and use what you feel is warranted for your facility. That's what I always did, and it worked well for me. Finally, when it comes to your Emergency Management Plan and Program, "know it", "live it", and "own it". After that, it's entirely up to you on just how much you can accomplish.

Over the course of this book, I've posed many questions to you, the Healthcare Emergency Manager. Some of these were answered within the text, while others were more rhetorical. The real purpose of many of these questions was to pique your interest and to stimulate dialogue and discussion on the points brought up. As I stated in the beginning of the book, this would not be a "how to" manual.

I don't have all of the answers, no one does. Each town, city, county, state and facility are different. I would never presume to try and tell some other professional in some other location how to do their job at their facility.

Rather, my intent, was to point you as a professional in your field, in a certain direction. A direction that will get you started asking questions you may not have considered and seeking different ways to prepare your facility. It will be your decision to choose the best way to accomplish this task. It is now up to you, to start asking those questions and seek out the answers you need. For if you do that successfully, you will have done your job by protecting your facility and staff against future disasters. It is my sincere hope that I have succeeded in my goal of getting you to think about those very questions. I wish you nothing but continued success in all of your future endeavors. But most of all, always stay safe.

## *Acknowledgements*

My wife Mary, my daughter Veronika and my son Mike for their unfailing love, support and encouragement during this entire project.

Ray Berke, author, friend and golf partner for his invaluable advice and guidance through the entire long book and publishing process.

George Araujo, truly an unsung hero, for his knowledge, expertise and friendship and for continuously providing me with current regulatory data.

# *Bibliography*

*Fundamentals of Healthcare Emergency Management* – U.S Department of Homeland Security, Center for Domestic Preparedness, July 2007

*Hospital Safety Exercises Toolkit* – Disaster Plans for all hazards & Joint Commission Compliance, Mary Russell, EdD, MSN, HCPro INC. 2008

*Mass Casualty Handbook: Hospital*- Emergency Preparedness & Response. 1st edition, Joseph A Barbera, MD Anthony G Macintyre, MD. Jane's Information Group, 2003

*Emergency Response Guidebook* – U.S. Department of Transportation 2012 edition.

*NFPA 1600 – Standard on Disaster Management & Business Continuity Programs.* National Fire Protection Association, 2013 edition, Quincy, MA.

*NFPA 99 – Healthcare Facilities Code.* National Fire Protection Association, 2012 edition, Quincy, MA.

*Planning & Protecting Medical/Hospital Infrastructure & Personnel.* Department of Homeland Security, Texas Engineering Extension Service, 2008.

*CBRN Response Handbook*- Antonio F. Garcia, Dan Rand, John Howard Rinard Jr., HIS Jane's Global Limited, 2011, 4th edition.

*The Joint Commission (TJC) Emergency Management Standards*- Oakbrook, Illinois, 2012- 2014.

**WMD Response Guidebook**- Louisiana State University (LSU) National Center for Biomedical Research, 2006, Version 3.3

**Emergency Preparedness Plans & Tools**; a Source List for Health Care Providers, Greater N.Y. Hospital Association.

**Effective Emergency Management Drills & Exercises**, Greater New York Hospital Association (GNYHA)

**29 CFR 1910.1200** – Hazard Communication Standard, Occupational Health & Safety Administration (OSHA) U.S. Government Printing Office.

**The Emergency Preparedness (Disaster) Plan in the Healthcare Facility** – Lee G. Shanley, Bachelor of Science Final Project, 1994.

www.ingramcontent.com/pod-product-compliance
Lightning Source LLC
Chambersburg PA
CBHW070317190526
45169CB00005B/1656